Dieter H. Marschall

Walther Pistols
Models 1 to PPX

Factory Variations and Copies
A Collector's and Historian's Overview
(Third, revised and enlarged English edition)

MOTT LAKE PUBLISHING

Dieter H. Marschall

Walther Pistols
Models 1 to PPX

Factory Variations and Copies
A Collector's and Historian's Overview
(third, revised and enlarged English edition)

Original German Edition:

Dieter H. Marschall
Walther Verteidigungspistolen
Mod. 1 bis PPX, Modellvarianten und Nachbauten, Dritte revidierte und erweiterte Auflage, dwj Verlags-GmbH, 74572 Blaufelden (Germany) 2014

ISBN 978-3-936632-11-8

Translation by Dieter H. Marschall
Proofreading by Heidi Cordes
Revision by Martin Klesper

Printed and bound in the USA

Published in the USA and other English speaking countries by
Mott Lake Publishing, P.O. Box 158294, Nashville, TN 37215

Publisher's Note Regarding Abbreviations in English Version

Readers of the English language version of this book will notice certain abbreviations with which they might not be well-acquainted. While we have provided a list of definitions below, please note that these abbreviations have been carried over directly from the original German language version of the book, even though the nuanced meaning of the original German version can sometimes be lost when translating to English.

cf. – "compare" or "compare to"
resp. – "respectively"
see there / see above / see before – reference to text or image located elsewhere in the book

English Version (Third)
brought to you by:

Mott Lake Publishing
www.MottLake.com

This book is dedicated to my twin grandchildren
Charlotte and Emilia

Contents

Foreword

When this book was first written in 1997 in German it was a product of irritation concerning these popular weapons in that there is so little reliable literature about them in German.

As the third German edition is sold out. I have in the meantime published the fourth, revised and enlarged edition. Basically this third English edition is textually identical with this fourth German one, as the corrections and additions are already worked in. Personally I no longer like some features of the first editions myself, so I have restructured and rewritten most of the text and included the new Walther models, too.

As I think most Walther collectors face the same problems when trying to classify Walther pistols as to their historical and technological background, I have collected this information to make it accessible to the public in an appropriate format.

This publication should serve as a stimulus to promote the discussion and further discovery of information on this subject.

Unfortunately the original documents of the Walther/Zella-Mehlis company have obviously been lost or were completely destroyed. The successor company, Walther/Ulm, has hardly any files dating 1945 or earlier.

This book about Walther pistols is supposed to help when classifying Walther models. The format chosen for the book is that of an identification sheet for each pistol model. This was done to present the information as compact as possible - hence the concise and not epic style of language. This book, therefore, is a "guide book", which in a similar form would e.g. be familiar to a botanist.

The sequence in which I have presented the different Walther models does not necessarily correspond chronologically with their introduction by the factory. First the blowback pistols are presented, then the pivoting locking block ones, then the later developments with the tipping barrels and last but not least the latest models.

Of course there are repetitions in the descriptions of the different models, if they show the same characteristics. Thus the readers need not read the whole book to find out the features of one specific model.

The very rare models, such as the Mod. KPK, Ultra or toggle lever action pistols, have not been included here. They are insignificant and financially unobtainable for most collectors.

So this book includes only Walther service or self-defense pistols. The target pistols, as well as the flare or signal pistols are not included here.

I have evaluated the currently available German literature. Hereby, I found that there are several contradictions. For example the statements in the Walther literature are often in contradiction to the data found in "Bundeswehr" (post-1945 German army) documents and regulations. Moreover, these often do not match with the information given by Walther/Ulm itself.

I have tried to resolve these contradictions by far reaching research and correspondence. Data, which do not appear well proven, are tagged with a question mark.

The production period of certain models – especially the beginning of a production – is often questionable. There are often mistakes or wrong conclusions in the Walther literature. It is pretty certain, however, that Walther did not launch a model onto the market before it was protected by patents – the more so if in the slide legends these patents were referred to.

The protection by a patent starts basically only with the granting and not yet with the application for a patent. The period of validity of any patent starts one day after the application. From 1877 to 1923 this period was 15 years, until 1978 it was 18 years. Only from 1978 onwards was this period of validity extended to 20 years. Before 1945 the so-called "Deutsche Reichs-Gebrauchsmuster" ("D.R.G.M."), often referred to as the "little patent", was valid for maximum ten years.

By the way it would have been a criminal offence, if Walther had referred to a patent in a slide inscription, which was not yet granted. Then it was only possible to write "Patent ang." (for "Patent angemeldet", i.e. patent pending).
Following the advice of Martin Krause, a renowned gun expert and patent attorney, I conclude that any pistol model could only be launched safely onto the market after the relevant patent was issued.

With regards to the contents of this book, I have mostly relied on the outstanding US-American Walther literature. Especially noteworthy here are the works of James Rankin, Prof. Warren Buxton, Gene Gangarosa and E.J. Hoffschmidt – see the bibliography in the appendix.

Furthermore I could consult my extensive databases containing information on Walther Mods. 1 to 9 and PPs respectively PPKs.

Further authors, whose books were consulted, are Manfred Kersten (Germany), Jan Balcar (Czech Republic), and Per Mathisen (Norway) - see bibliography.

I would like to thank Mr. Werner Hampel, Mr. Joachim Hertrampf and Mr. Manfred Wörz (all from Walther/Ulm), as well as Mr. Heinrich (Wehrtechnische Studiensammlung, Koblenz). To them I am very grateful for information and photographs.

In addition I have obtained much valuable information through correspondence with collector colleagues. Among these are (in alphabetical order): Don Anderson, Jim Cate, John Massaro, Dave Shike, Steven J. Stepan, Doug Wirt, Joe Wotka (all USA), Burkhard Fettinger, Josef Mötz (both Austria), Jürg Siegenthaler, Philippe Balogh (both Switzerland), Bernd Dahinten, Hendrik Frühauf, Joachim Görtz, Dietrich Kersten, Martin Krause, Alexander Krutzek, Theo Löffler, Dr. Richard Preuß, David F. Reichelt, Wolf-Dietrich Roth, Albert Stadler, Egon Thiel, Wolfgang Trunschke, Gregor Wensing and Sigurd Wypich (all Germany). To all of them a heartful thank you.

My special thanks go to my "Walther mentor", Karl Schäfer, who has published several articles on Walther pistols himself – see bibliography. Without his energetic help and his friendly encouragement the first German edition of this book would not even have been started in 1995.

Further thanks go to my two colleagues: Manfred Kersten, who helped me with information and other material and Martin Klesper, who not only contributed information but revised this translation with the critical eye of a technician.

After the sale of the publishing house "DWJ-

Verlag" all the printing files and pictures of the first and second German editions were lost. So I would not even have considered a third German edition without the active help of Wilfried Garbers (Germany), Dr. Geoffrey Sturgess (formerly UK, now Switzerland) and Bob Adams (USA). They helped me with all the new pictures. Special thanks to Wilfried Garbers for most of the pictures and a load of impulses. Furthermore he has proofread the manuscript and contributed with his knowledge of the subject.

The same is true for Cord Rather, who read this manuscript critically and contributed with his knowledge.

I want to thank my wife Gabriele, who patiently endured my activities and even typed some parts of the manuscript.

As I know that many fellow collectors have an extensive knowledge of this subject, but are sometimes a bit reluctant to share their information, I herewith welcome any contributions, corrections and impulses. They are invited to do so per email to me (dhimarshall@gmx.de) or by letter to the publisher.

Summer 2016
Dieter H. Marschall

Preliminary remarks on serial number ranges:

The serial numbers (short SN) given in this book are based on the extensive databases of the author. The delimitation of the serial ranges of different models cannot be very sharp, as there were always transitional phases, where Walther used up older parts when producing newer models and variations. This was mostly done out of economic reasons. So most statements on serial numbers can only be approximate.

Walther Mod. 1

Picture 1: *Walther Mod. 1, 2nd variation, SN 29 345, caliber 6.35 mm Browning (.25 ACP), "two buttons variation"*

Production:	1910 - 1915 (?)
Caliber:	6.35 mm Browning = .25 ACP (6 rounds)
Total Length:	4.5 in.
Barrel Length:	2.0 in.
Weight:	13.1 oz.
Rifling:	4 or 6 grooves, right-twisted
Operation:	hammerless, unlocked blowback action, striker-fired

Preliminary remarks on the production period:

In early catalogues the "Mod. 1" is often designated as "Mod. 1900". The first application for a patent for a "blowback operated pistol with a fixed barrel" was filed on 7th August 1909. The relevant patent, however, was granted on 22nd June 1911. Thus the Walther Mod. 1 was obviously produced no earlier than 1910 – then with the additional designation "DRGM In- & Auslandspatente ang." ("Deutsche Reichs-Gebrauchsmuster", National and Foreign Patents Pending) in the slide inscription. Only after the patent was granted late in June 1911 could be this changed to "D.R.P & Auslandspatente" ("Deutsches Reichs Patent" and Foreign Patents) or simply "Patent" only.

The US patent 991 398 was filed on 9th June 1910 and granted on 2nd May 1911.

The patent 54 841 for Austria was filed on 23rd May 1910.

Early Mod. 1 pistols still have the proof mark "Crown over U" – used by the Zella-Mehlis proof house up to 9/1911. Later Mod. 1 pistols have "Crown over N" proof mark – used from 9/1911 onwards.

Mod. 1

Characteristics:

- Stepped slide, which embraces the barrel sleeve only partially
- Fixed barrel
- Barrel sleeve with front sight on top
- No rear sight, only sighting groove
- Top ejection
- Recoil spring under barrel
- Safety button as push button located traverse in the frame behind the grip plates
- Trigger bar only partially visible
- Disconnector: first one-piece, later two-piece
- Bottom magazine release
- Metal finish: salt blued
- Grip plates made of black, checkered, hard rubber with intertwined "CW" logo (for "Carl Walther") on both sides or "6,35" on left hand side, ivory grips at an extra charge
- Some Mod. 1 pistols have longer barrels e.g. for export to Austria-Hungary to comply with gun laws there.

Mod. 1, 1st Variation (so-called "one button variation")

Characteristics:

- Slide legend left hand side:
- Up to June 1911: "Walther's Selbstlade-Pistole Cal. 6,35 D.R.G.M. In-& Auslandspatent ang."
- Or: "Selbstlade-Pistole Cal. 6,35 Walther's Patent"
- Or: "Selbstlade-Pistole Cal. 6,35 Patent"
- Inscription with or without Walther banner (in early, less curved form)
- Proof marks from 1910: "Crown over U", from September 1911: "Crown over N"
- Inscription on frame of some early Mod. 1 pistols: "Made in Germany" (behind the grip plates)
- Only one button in frame (safety push button)
- Recess around the trigger in a rounded shape

SN 4 to 3 500

- Proof marks "Crown over U" (1910 – 9/1911)
- Some very early specimens (e.g. SN 17) have a shorter frame, barrel and slide at the front and no logos on the grip plates.
- 13, 15 or 17 slide serrations on both sides of slide (With very early specimens the 15 slide serrations are still straight and running through to the lower edge of the slide)
- Two-piece disconnector
- Straight trigger bar
- Magazine butt plate nearly straight at the front

SN 3 500 to 8 000: as before, but

- Proof marks "Crown over N" (from 9/1911)
- 13 or 17 slide serrations

SN 8 000 to 11 300: as before, but

- 12 slide serrations
- From about SN 10 000: stepped trigger bar

SN 11 300 to 15 500: as before, but

- 13 slide serrations
- From about SN 13 000: one-piece disconnector

Mod. 1, 2nd Variation (so-called "two button variation")

Characteristics:

- Second push button on the right side of the frame (behind the safety button), which serves as a dismantling device
- Slide legend left hand side: "Selbstlade-Pistole Cal. 6,35 Walther's Patent" or

"Selbstlade-Pistole Cal. 6,35 / Patent", resp. "Walther's Selbstlade-Pistole Cal. 6,35 / D.R.P. & Auslandspatente" – with or without Walther banner

SN 15 500 to 31 000

- 12 or 13 slide serrations
- Stepped trigger bar
- One-piece disconnector
- Recess around the trigger in a more rectangular shape
- Magazine butt plate rounded at the front
- From about SN 21 000: Walther banner in later, more curved form

Total production of Walther Mod. 1 pistols: about 30,000 pistols

Walther Mod. 2

Picture 2: *Walther Mod. 2, SN 8407, caliber 6.35 mm Browning (.25 ACP), variation without loaded chamber indicator*

Production:	1913 - 1915 (?)
Caliber:	6.35 mm Browning = .25 ACP (6 rounds)
Total Length:	4.2 in.
Barrel Length:	2.1 in.
Weight:	9.75 oz.
Rifling:	4 or 6 grooves, right-twisted
Operation:	unlocked blowback action with shrouded hammer and firing pin

Preliminary remarks:

With this pistol Mod. 2 Walther started a new generation of pistol models, which ended with Mod. 7. Typical of these models are the slide with its closed top, the shrouded hammer and the pivoting safety lever behind the left grip plate.

On 22nd October 1911 the patent for a "blowback pistol with a fixed barrel and a shell-type front slide" was filed. This patent was later realized in the pistols Mod. 2 to Mod. 7. The patent number 256 606 referring to this application was granted as late as 17th February 1913 and so the pistol Mod. 2 could be sold from this date onwards.

The pistol Mod. 2 is often confused with the later and almost identical Mod. 5. Reliable discerning features are, however, the location of the serial number, the number of slide serrations as well as the place name in the slide inscription.

The first and the second variation of the pistol Mod. 2 did not follow each other chronologically, but were contemporaneous variations. The one with the loaded chamber indicator ("automatic rear sight") was an option at an extra charge. Here a spring operated rear sight is lifted out of the slide, if a round is fed to the chamber.

The Swiss Walther expert Jürg Siegenthaler is of a different opinion as to the sequence of the two variations (see chapter Mod. 5).

Picture 3: *Walther Mod. 2, SN 2749, caliber 6.35 mm Browning (.25 ACP), variation with loaded chamber indicator*

Characteristics:

- Slide legend left hand side: "Selbstlade-Pistole Cal. 6,35 Walther's Patent" with Walther banner (caliber rarely given in inches ".25")
- Slide legend right hand side: first without company address, from about SN 27 000: "Carl Walther Waffenfabrik Zella-St. Blasii"
- Proof marks: "Crown over N"
- Serial number horizontally on right hand side of trigger guard
- Fixed barrel, which serves as a guide for the recoil spring
- Barrel not reinforced – cf. late variation of Mod. 5
- Knurled muzzle nut, which serves as a retainer for the slide and the recoil spring (cf. patent number 256 606 from 17th February 1913 – see above)
- No rear sight, only sighting groove
- 9 slide serrations on both sides of the slide
- Ejection port on right side of slide
- Shrouded hammer

- Internal trigger bar
- Bottom magazine release
- Pivoting safety lever behind left grip plate
- Magazine safety (early pistols only)
- Grip plates made of black, checkered, hard rubber (mostly with intertwined "CW" for Carl Walther, rarely with "Caliber 6,35"), wooden or ivory grips optional at an extra charge
- Metal finish: salt blued, nickel-plated finish optional at an extra charge

Mod. 2, Variation with loaded chamber indicator

SN 1000 to 20 000 (from early 1914?)

Characteristics:

- "automatic rear sight" with front sight inserted into the rib sideways;
- Patent number 271 863 referring to this feature was filed on 8th January 1913, granted on 19th March 1914 – hence a production before the beginning of 1914 is improbable
- Serial number on some pistols additionally on the left side of the slide (e.g. with SN 15 134)

Mod. 2, Variation without loaded chamber indicator

SN 1770 to 71 000 (serial number range overlaps)

Characteristics:

- First sighting groove with bead front sight only, later fixed rear sight
- From about SN 27 000 company address on the right side of the slide

Special models resp. Markings:

- Misspellings in company address: e.g. SN 25 570 inscribed with "Zella-St. Basii" instead of "St. Blasii"
- Model designation "Mod. 2" (on left hand side of slide, e.g. with SN 13 988): not authentic, but engraved later in the USA
- Some Mod. 2 pistols have longer barrels e.g. for the export to Austria-Hungary to comply with gun laws there

Total production of Walther Mod. 2 pistols: about 19,000 pistols

Walther Mod. 3

Production:	1913 - 1915 (?)
Caliber:	7.65 mm Browning = .32 ACP (6 rounds)
Total Length:	5.0 in.
Barrel Length:	2.64 in.
Weight:	6.5 oz.
Rifling:	6 grooves, right-twisted
Operation:	unlocked blowback action with shrouded hammer and firing pin

Preliminary remarks:

On 22nd October 1911 a patent for a "blowback pistol with a fixed barrel and a shell-type front slide" was applied for. This patent was later realized in the pistols Mod. 2 to Mod. 7. The patent number 256 606 referring to this application was granted as late as 17th February 1913 and so the pistol Mod. 3 could be sold from this date onwards.

Mod. 3

Characteristics:

- Slide legend left hand side: "Selbstlade-Pistole Cal. 7,65 Walther's Patent" first without, then with Walther banner

- Slide legend right hand side: none
- Proof marks: "Crown over N"
- Serial number vertically on right hand side of frame
- Fixed barrel, which serves as a guide for the recoil spring
- Muzzle nut, which serves as a retainer for the slide and the recoil spring (cf. patent number 256 606 from 17th February 1913 – see above)
- No rear sight, only sighting groove; front sight on top of muzzle nut
- 12 or 13 vertical slide serrations on both sides of the slide
- Ejection port on left side of slide
- Shrouded hammer
- Partially visible, stepped trigger bar
- Bottom magazine release
- Pivoting safety lever behind left grip plate
- Disassembly lever below the front of the slide
- Grip plates made of black, checkered, hard rubber (mostly with intertwined "CW" for "Carl Walther", rarely with "Caliber 7,65"), ivory grips optional at an extra charge
- Metal finish: salt blued, nickel-plated finish optional at an extra charge

Mod. 3, 1st Variation

SN 11 000 to 30 000 (SN range overlaps with Mod. 3, 2nd variation)

Characteristics:

- Mostly 13 sometimes 12 vertical slide serrations
- Slide legend on the left side of the slide without Walther banner

Mod. 3, 2nd Variation

SN 21 800 to 35 000 (SN range overlaps with Mod. 4, 1st variation)

Characteristics:

- Mostly 12 sometimes 13 vertical slide serrations
- Slide legend on the left side of the slide with Walther banner

Picture 4: *Walther Mod. 3, second variation, SN 26 599, caliber 7.65 mm Browning (.32 ACP)*

Special models resp. Markings:

- SN 25 158: left side of slide inscribed with "Globe Arms Co. Germany", stamped with "Krupp Steel" in ejection port – provenance unknown
- SN 25 214: left side of slide inscribed with "Messerfabrik Vulkan GmbH, Solingen Germany", right side of slide with "Tenson registered trademark" – provenance: possibly a licensed production for export purposes? (Actually "Tenson" is a Swedish trademark for outdoor clothing)
- Some Mod. 3 pistols have longer barrels e.g. for the export to Austria-Hungary to comply with gun laws there (e.g. SN 23 615).

Total production of Walther Mod. 3 pistols: about 3,500 pistols

Walther Mod. 4

Production:	1913 - 1928/29 (?)
Caliber:	7.65 mm Browning = .32 ACP (6 rounds)
Total Length:	6.0 in.
Barrel Length:	3.3 in.
Weight:	19.4 oz.
Rifling:	4 or 6 grooves, right-twisted
Operation:	unlocked blowback action with shrouded hammer and firing pin

Preliminary remarks:

The Walther pistol Mod. 4 was protected by a patent, too. Hence it carries "Walther's Patent" in its slide inscription. On 22nd October 1911 a patent for a "blowback pistol with a fixed barrel and a shell-type front slide" was filed. This patent was later realized in the pistols Mod. 2 to Mod. 7. The patent number 256 606 referring to this application was granted as late as 17th February 1913 and so the pistol Mod. 4 could be sold from this date onwards.

Throughout the production period of the Mod. 4 there were several technical changes. The variations differ in the sights, the number of slide serrations or the type of slide legend only.

Mod. 4

Characteristics: like Mod. 2 and Mod. 3

- Slide legend left hand side: varies (see below)
- Slide legend right hand side: varies (see below) or is non-existent (see licensed production)
- Proof marks: "Crown over N"
- Military acceptance marks (if issued to the military): "Crown over gothic letter" (on left hand side of slide)
- Serial number vertically on right hand side of frame
- Fixed barrel, which serves as a guide for the recoil spring
- Muzzle nut, which serves as a retainer for the slide and the recoil spring (cf. patent number 256 606 from 17th February 1913 – see above); front sight on top of muzzle nut
- Some Mod. 4s without a rear sight (sighting groove only), some with adjustable rear sight
- 7, 12 or 16 slide serrations on both sides of the slide
- Ejection port on left hand side of slide
- Shrouded hammer

- Trigger bar first partially visible and stepped, later internal
- Bottom magazine release
- Pivoting safety lever behind left grip plate
- Grip plates made of black, checkered, hard rubber (mostly with intertwined "CW" for "Carl Walther", some early Mod. 4s with "Caliber 7,65"), horn grips optional at an extra charge
- Metal finish: salt blued, nickel-plated finish optional at an extra charge

Mod. 4, 1st Variation (1st transitional Model 3/4)

SN 20 000 to 33 940

Characteristics:

- Slide legend left hand side: "Selbstlade-Pistole Cal. 7,65 Walther's Patent"
- Slide legend right hand side: "Carl Walther Waffenfabrik Zella-St. Blasii"
- 9 or 12 slide serrations (mostly vertical) – cf. Mod. 3
- Disassembly lever below the front of the slide – cf. Mod.3
- Trigger bar partially visible
- Top of slide rounded, front of slide bottom square – cf. Mod. 3
- Sights: only sighting groove with front sight on muzzle nut

Picture 5: *Walther Mod. 3/4, 2nd transitional model, SN 34 158, caliber 7.65 mm Browning, (.32 ACP)*

Mod. 4, 2nd Variation (2nd transitional Model 3/4)

SN 33 900 to 48 000 (SN range overlaps with 1st variation)

Characteristics:

- 7 slide serrations (no longer vertical, but slightly angled to the front)

Mod. 4, 3rd Variation

SN 48 000 to 68 700

Characteristics: as before, but

- No disassembly lever below the front of the slide – cf. Mod.3
- 7 or 9 slide serrations
- Mod. 4s made under license (see below): no slide inscription on right hand side

Mod. 4, 4th Variation

SN 67 700 to 223 000
(SN range overlaps with 3rd variation)

Characteristics: as before, but

- Sights: mostly adjustable rear sight, front sight on muzzle nut; rarely sighting groove only
- From SN 210 000 onwards some slide legends on right hand side with the new designation "Zella-Mehlis I" instead of "Zella-St. Blasii"

Remarks:

Officially the merging of the two towns "Zella-St. Blasii" and "Mehlis" to the twin town "Zella-Mehlis" was carried out as late as 1919 during a general municipal reform. The new place name "Zella-Mehlis", however, was used unofficially in company addresses or in other usages from 1893 already (e.g. in the name of the "Herzogliche Beschussanstalt Zella-Mehlis"). Walther seems to have used the new place name "Zella-Mehlis" partially from 1916 onwards. Often the name "Zella-Mehlis I" was then used for the old "Zella-St. Blasii".

Picture 6: *Walther Mod. 4, 4th variation, SN 211 813, caliber 7.65 mm Browning (.32 ACP)*

Mod. 4, 5th Variation

SN 223 000 to 257 000

Characteristics: as before, but

- Slide legend left hand side: "Walther's Patent Cal. 7,65" (in italics)
- Slide legend right hand side: "Waffenfabrik Walther Zella-Mehlis" (in italics)
- 16 slide serrations
- Internal trigger bar
- Adjustable rear sight, front sight on muzzle nut
- Up to about SN 224 700 ejection port rounded, later mostly more rectangular

Mod. 4, 6th Variation

SN 257 000 to 288 800 and 442 300 to 493 700

Remark:

Walther had to reorganize their serial number system – presumably because of the later introduction of the PP and PPK models.

Characteristics: as before, but

- Slide legend right hand side: "Waffenfabrik Walther Zella-Mehlis (Thür.)" (in italics)
- Rectangular ejection port only

Special models resp. Markings:

- All SN ranges: stamp "Made in Germany" on pistols for export purposes
- SN 26 960 (early transitional model 3/4): "Made in Germany" (on left hand side of slide instead of Walther banner), no inscription on right hand side of slide (!)
- SN 256 163: inscription "E Kal. 7,65" (on left hand side of slide) and longer slide, i.e. presumably an experimental pistol in caliber 6.45 mm Ultra (bottleneck cartridge) – despite the differing caliber designation
- SN 487 419: Czech proof mark dated 1925
- SN range 121 000 to 125 100: some pistols with misspelling ("Zela" instead of "Zella")
- SN range 263 800 to 264 200: Some pistols carry the company name "Heinrich Krieghoff" (additionally stamped on left hand side of slide). The Walther company address the right hand side the of slide is stamped over. The reasons for this are unknown.
- SN range 482 900 to 483 150: Some pistols carry the stamp "D.R." (with wings) on the slide. Those Mod. 4s were issued to the railway police of "Deutsche Reichsbahn"
- Stamp "LW" (mostly on front gripstrap): Those Mod. 4s were issued to the Austrian "Landwehr" (territorial army)
- Stamp "ByP": Mod. 4s which were reused by the Bavarian police after 1945

Licensed production

SN 60 000 to 99 500 (3rd and 4th variation) and **133 500 to 150 000** (4th variation)

At the beginning of World War I the standard service pistol of the German military was the Luger pistol in caliber 9 mm Luger – introduced in the army as "Pistole P08" and in the navy as "Pistole P04". As the army did not have enough Luger pistols, however, pistols in caliber 7,65 mm Browning (.32 ACP) were introduced as so-called "Behelfspistolen" (provisional pistols). In the "Armee-Verordnungsblatt Nr. 40" (army ordinance sheet) from 11th August 1917 a great number of pistols is named for the sale to officers – amongst them the "Walther-Pistole 7,65 mm".

However, at the beginning of World War I Walther was unable to meet the demand of 30 000 pistols Mod. 4 of the "Heeresverwaltung" (army administration) alone. So on 4th December 1915 Carl Walther signed a license contract with Immanuel Meffert/Suhl for the licensed production of such pistols.

Pistols Mod. 4 made under license do not carry the Walther designation on the right side of the slide, but the mark of the licensee on the trigger guard.

The volume of this contract was too big for the capacities of Meffert, too. So they already arranged in the license contract with Walther that they could use subcontractors (named "Unterlieferanten" in the contract) to meet the demands of the military.

The following Thuringian companies, including Meffert, were named as licensees. The different marks of the companies are given in brackets. Those marks were stamped onto the left hand side of the trigger guard (except for the "M" for "Mercedes, which was stamped on the right hand side):

Licensee:

- Immanuel Meffert, Suhl ("J.M.")

Subcontractors:

- Mercedes Bureaumaschinen, Mehlis ("M" in an oval)
- Heinrich Krieghoff, Suhl (stylized anchor) – only production of frames?
- F.W. Kessler, Suhl ("F.W.K.")
- Schmidt & Habermann, Suhl ("S&H")
- Gebrüder Rempt, Suhl ("GR")
- Christoph Funk, Suhl ("F")
- August Schüler, Suhl ("AS")
- Gebrüder Merkel, Suhl ("GM" in a shield)
- Sempert & Krieghoff, Suhl (only production of frames for other licensees)
- H.M. Gering, Heidersbach ("G")
- Wilhelm Weiss, Heidersbach (only production of slides and rear sights)

In these contracts Meffert and the subcontractors were obliged to produce at least 30 000 pistols Mod. 4 in between December 1915 and March 1917.

In the subcontracts there are some contradictions, however: The contract dated 18th February 1916 between Meffert and the subcontractors names a total number of 11 025 pistols to be delivered with the following subdivision:

Immanuel Meffert:	32 % (3 528 pistols)
F.W. Kessler:	16 % (1 764 pistols)
Schmidt & Habermann:	16 % (1 764 pistols)
Gebrüder Rempt:	16 % (1 764 pistols)
August Schüler:	8 % (882 pistols)
Christoph Funk:	12 % (1 323 pistols)

It was agreed moreover that the company "Concern Merkel" deliver 7 950 pistols. Mercedes Bureaumaschinen should contribute 11 025 pistols to the total of 30 000 pistols. In that contract Meffert and Mercedes are subsumed under the designation "Concern Meffert".

In the subcontract dated 20th November 1916 the company Mecedes is no longer named, however. In the second delivery of Mod. 4s there are no pistols known, which show the "M" on the trigger guard.

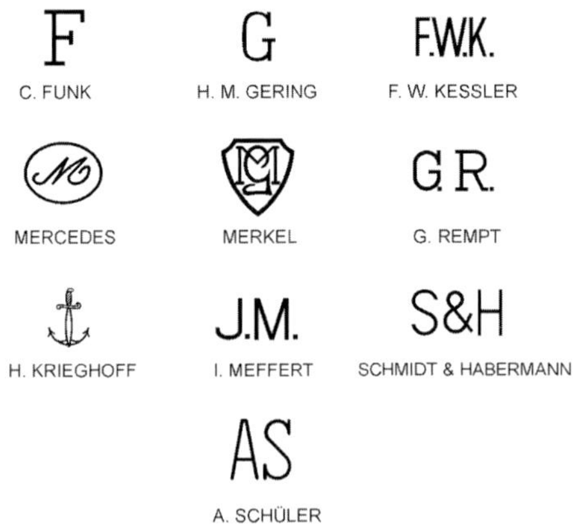

Subcontractor markings for the Walther Model 4. **Source**: *Jan Balcar, Pilsen*

The contractors agreed on a total production of 480 pistols per week. This production was subdivided as follows:

Immanuel Meffert:	113 pistols
F.W. Kessler:	60 pistols
Schmidt & Habermann:	60 pistols
Gebrüder Rempt:	60 pistols
Gebrüder Merkel:	60 pistols
August Schüler:	25 pistols
Christoph Funk:	40 pistols
H.M. Gering	32 pistols
H. Krieghoff	30 pistols (plus 450 frames)

It was agreed that Krieghoff produce 480 frames per week – 30 for themselves and 450 frames (with barrels already pressed in) for the other subcontractors. Gebrüder Merkel were in charge of the delivery of minor parts (slides, magazines, rear sights, etc.). Those rear sights again were delivered to Krieghoff by Wilhelm Weiss, Heidersbach. Thus there was a consortium created in which most of the participating companies were limited to the assembly of parts delivered by other companies.

Within almost 15 months a total of 30 000 Mod. 4s should thus be produced under license.

Oddly enough a company named "Venus Waffen-Werk" (Oscar Will) from Zella-Mehlis tried to make the Bavarian military place an order for "Walther Selbstladepistolen Kaliber 7,65 (Mod. 4)". Although this company produced mainly airguns, they managed to get this order. In the end this contract did not make them completely happy though. In March 1915 Oscar Will delivered four sample pistols to "Gewehrfabrik Amberg" and until 17th July 1915 further 1500 Mod. 4s with ammunition and holsters to the Bavarian "Feldzeugmeisterei" (army ordnance depot). One year later the ammunition, which was delivered together with the pistols, proved to be inappropriate for the pistols. Despite the chaos of war the remaining 86 pistols were collected and checked. The result was that there were misfeedings due to wrongly dimensioned chambers and magazines. One of the reasons for this problem was that Oscar Will had delivered Mod. 4s, which were not dimensionally accurate (second choice?). Another reason, however, was that the ammunition did not meet the standards.

Use of Walther Pistols Mod. 4

Officers and high-ranking military employees ("Militärbeamte") had to procure their service pistols privately. For them the Walther Mod. 4 was a popular model. Officially the Mod. 4 was only a "Behelfswaffe" (provisional weapon), however.

The Prussian "Kriegsministerium" shipped 72 000 Mod. 4 pistols to the state of Bavaria via "Königliche Gewehrfabrik Erfurt".

Some police forces (including the railway police) introduced the Walther Mod. 4 as a service pistol – amongst them the police of the state of Hessia (e.g. SN 256 914 and SN 256 929 marked with "H.P." on the front gripstrap) or the police of the Danish capital Kobenhavn (marked "Kbh. Pol. Nr ....).

The Finnish army introduced the Walther Mod. 4 as a service pistol with the military designation "Pistooli M/Walther 4". As late as June 1971 there were still 560 Mod. 4s in the Finnish depots.

Total production of Walther Mod. 4 pistols (including licensed production): about 351,000 pistols (about 90 000 of those for the military)

Walther Mod. 5

Period:	1915 - 1923 (?)
Caliber:	6.35 mm Browning = .25 ACP (6 rounds)
Total Length:	4.2 in.
Barrel Length:	2.1 in.
Weight:	9.71 oz.
Rifling:	4 or 6 grooves, right-twisted
Operation:	unlocked blowback action with shrouded hammer and firing pin

Preliminary remarks:

The Walther pistol Mod. 5 was protected by a patent, too. Hence it carried "Walther's Patent" in its slide inscription. On 22nd October 1911 a patent for a "blowback pistol with a fixed barrel and a shell-type front slide" was filed. This patent was later realized in the pistols Mod. 2 to Mod. 7. The patent number 256 606 referring to this application was granted as late as 17th February 1913.

The pistol Mod. 5 is often confused with the earlier and almost identical Mod. 2. Reliable distinctive features are, however, the location of the serial number, the number of slide serrations as well as the place name in the slide inscription.

The serial number ranges of Mod. 2 and Mod. 5 partially overlap. As the product quality of the Mod. 5 pistols is better than the one of the Mod. 2s, it was perhaps only a marketing trick to give the improved model a new name.

The Swiss author Jürg Siegenthaler wrote in 2007 (see bibliography) that the difference between the two models is only a question of the type of rear sight. All pistols with the "automatic rear sight" (loaded chamber indicator), guilloché slide rib and magazine safety are Mod. 2s – no matter how many slide serrations there are or where the serial number is located. According to his article the Walther Mod. 5 is only a cheaper, but basically identical version of the Mod. 2 pistol – without the costly "automatic rear sight".

Siegenthaler classifies as follows:

- Early variation: 9 slide serrations, sighting groove, serial number either right hand side on the trigger guard or left hand side on the frame
- Middle variation: 16 slide serrations, sighting groove, serial number on left hand side of the frame
- Late variation: 16 slide serrations, guilloché slide rib with rear and front sights, serial number on left hand side of the frame

Siegenthaler's classification is mostly based on the texts and illustrations in Walther brochures.

Officially the merging of the two towns "Zella-St. Blasii" and "Mehlis" to the twin town "Zella-Mehlis" was carried out as late as 1919 during a general municipal reform. The new place name "Zella-Mehlis, however, was used unofficially in company addresses or in other usages from 1893 already (e.g. in the name of the "Herzogliche Beschussanstalt Zella-Mehlis"). Walther seems to have used the new place name "Zella-Mehlis" partially from 1916 onwards. Often the name "Zella-Mehlis I" was then used for the old "Zella-St. Blasii" – cf. Mod. 4.

Mod. 5

Characteristics: (similar to Mod. 2)

- Slide legend left hand side: varies (see below)
- Slide legend right hand side: varies (see below)
- Proof marks: "Crown over N"
- Serial number vertically on left hand side of frame (rarely on right hand side)
- Fixed barrel, which serves as a guide for the recoil spring
- Early Mod. 5s with non-reinforcd barrel, later specimens with reinforced barrel

Picture 7: *Walther Mod. 5, transitional Model 2/5, SN 79 641, caliber 6.35 mm Browning (.25 ACP)*

(outer diameter non-reinforced: 8.0 – 8.5 mm, reinforced 9.0 – 9.5 mm)

- Knurled muzzle nut, which serves as a retainer for the slide and the recoil spring (cf. patent number 256 606 from 17th February 1913 – see above)
- Some Mod. 5s with a sighting groove only, some with a milled front and rear sight
- 9 or 16 slide serrations on both sides of the slide
- Ejection port on right hand side of slide
- Shrouded hammer
- Internal trigger bar
- Bottom magazine release
- Pivoting safety lever behind left grip plate
- No magazine safety
- Grip plates made of black, checkered, hard rubber (with intertwined "CW" for "Carl Walther"), horn or ivory grips optional at an extra charge
- Metal finish: salt blued, nickel-plated finish optional at an extra charge
- Only very few engraved Mod. 5s are known

Mod. 5, 1st Variation (transitional Model 2/5)

SN 63 000 to 83 540

Characteristics:

- Slide legend left hand side: "Selbstlade-Pistole Cal. 6,35 Walther's Patent" with Walther banner – cf. Mod. 2
- Slide legend right hand side: "Carl Walther Waffenfabrik Zella-St. Blasii" or (from 1916 – cf. Mod. 4): "Carl Walther Waffenfabrik Zella-Mehlis I"
- Barrel not reinforced
- Sights: only sighting groove with bead front side in sighting groove

- Mostly 9 slide serrations
- Ejection port in rounded shape

Mod. 5, 2nd Variation

SN 83 500 to 93 000 (SN range overlaps with 1st variation)

Characteristics: as before, but

- Slide legend left hand side: "Walther's Patent Cal. 6,35" with Walther banner
- Slide legend right hand side: "Waffenfabrik Walther Zella-Mehlis"
- Barrel reinforced
- Sights: slide rib with milled front and rear sight
- 16 slide serrations

Mod. 5, 3rd Variation

SN 86 200 to 102 000 (SN range overlaps with 2nd variation) and

SN 370 000 to 390 000 (There is a gap in the serial number range due to a reorganization of the serial number system)

Characteristics: as before, but

- Ejection port in more rectangular shape
- Serial number sometimes vertical on right hand side of frame

Special models resp. Markings:

- Stamp "Made in Germany" on specimens for export purposes
- Some Mod. 5s show serial numbers with three or four digits only. The reasons for this are unknown. Maybe these serial numbers belong to Mod. 2 pistols.

Total production of Walther Mod. 5 pistols: about 120,000 pistols

Walther Mod. 6

Production:	1915 (?) - 1917
Caliber:	9 mm Luger = 9 x 19 mm (8 rounds)
Total Length:	8.3 in.
Barrel Length:	4.75 in.
Weight:	34 oz.
Rifling:	4 grooves, right-twisted
Operation:	unlocked blowback action (despite the powerful caliber!) with shrouded hammer and firing pin

Preliminary remarks:

The Walther pistol Mod. 6 is basically an enlarged Mod. 2 – 5, with ejection to the right hand side. With this new model Walther tried to develop a service pistol for the armed forces in the standard military caliber 9 mm Luger.

The military refused the Mod. 6, however, due to the weak design of the unlocked blowback action, which could not handle the recoil of the powerful 9 mm Luger cartridge. Moreover enough P08 pistols and so-called "Behelfswaffen" (provisional weapons) were available then.

So only a few Mod. 6s were issued as "Behelfswaffen" to the armed forces. Those have the military acceptance stamp ("Crown over gothic letter) on the left hand side of the slide. Some officers bought their Mod. 6s on the commercial market, however, and carried them as their service pistol.

Some of the Mod. 6s were exported to foreign markets. Those were stamped "Made in Germany".

Picture 8: *Walther Mod. 6, SN 734, caliber 9 mm Luger*

Mod. 6

SN 1 to 1038

Characteristics: (similar to Mod. 4)

- Slide legend left hand side: "Selbstlade-Pistole Cal. 9 m/m Walther's Patent"
- Slide legend right hand side: "Walther Waffenfabrik Zella-St. Bl." (although from about 1916 the new name "Zella-Mehlis I" was partially used already – cf. Mod. 4 and 5)
- Proof marks: "Crown over N"
- Serial number horizontally on right hand side of trigger guard
- Fixed barrel, which serves as a guide for the recoil spring
- Grooved muzzle nut, which serves as a retainer for the slide and the recoil spring (cf. patent number 256 606 from 17th February 1913 – see above)
- Front sight on muzzle nut, adjustable rear sight
- 7 slide serrations on both sides of the slide
- Ejection port on right hand side of slide
- Shrouded hammer
- Internal trigger bar
- Bottom magazine release
- Pivoting safety lever behind left grip plate
- No magazine safety
- Grip plates made of black, checkered, hard rubber (with intertwined "CW" for "Carl Walther")
- Metal finish: salt blued

Total production of Walther Mod. 6 pistols: about 1,000 pistols

Walther Mod. 7

Picture 9: *Walther Mod. 7, 1st variation, SN 29 066, caliber 6.35 mm Browning (.25 ACP)*

Production:	1915 - 1919 (?)
Caliber:	6.35 mm Browning = .25 ACP (8 rounds)
Total Length:	5.3 in.
Barrel Length:	3.0 in.
Weight:	12 oz.
Rifling:	6 grooves, right-twisted
Operation:	unlocked blowback action with shrouded hammer and firing pin

Preliminary remarks:

The Walther pistol Mod. 7 is basically a downsized Mod. 6 – with the ejection port on the right hand side, too.

The variations differ in the number of slide serrations or in the slide legend only.

Mod. 7

Characteristics:

- Slide legend left hand side: "Selbstlade-Pistole Cal. 6,35 Walther's Patent" with Walther banner
- Slide legend right hand side: varies (see below)
- Proof marks: "Crown over N"
- Serial number vertically on left hand side of frame or horizontally on right hand side of the trigger guard
- Fixed barrel, which serves as a guide for the recoil spring
- Muzzle nut, which serves as a retainer for the slide and the recoil spring (cf. patent number 256 606 from 17th February 1913 – see above)
- Sights: front sight on top of muzzle nut, fixed rear sight with sighting groove

- 9 or 16 slide serrations on both sides of the slide
- Ejection port on left hand side of slide
- Shrouded hammer
- Internal trigger bar
- Bottom magazine release
- Pivoting safety lever behind left grip plate
- Grip plates made of black, checkered, hard rubber (mostly with intertwined "CW" for "Carl Walther", some early Mod. 7s with "Caliber 6,35"), horn grips optional at an extra charge
- Metal finish: salt blued, nickel-plated finish optional at an extra charge

Mod. 7, 1st Variation

SN 4 000 to 5 500 and 27 500 to 60 000
(There is a gap in the serial number range due to a reorganization of the serial number system)

Characteristics:

- Slide legend left hand side: "Selbstlade-Pistole Cal. 6,35 Walther's Patent" with Walther banner
- Slide legend right hand side: "Carl Walther Waffenfabrik Zella-St. Blasii"
- 9 slide serrations
- Some specimens with the serial number horizontally on right hand side of trigger guard

Mod. 7, 2nd Variation

SN 60 000 to 65 500 (from 1916?)

Characteristics: as before, but

- Slide legend left hand side: "Selbstlade-Pistole Cal. 6,35 Walther's Patent" with Walther banner
- Slide legend right hand side: "Carl Walther Waffenfabrik Zella-Mehlis I"
- 16 slide serrations

Mod. 7, 3rd Variation

SN 64 500 to 84 000

Characteristics: as before, but

- Slide legend left hand side (in italics): "Walther's Patent Cal. 6,35" with Walther banner
- Slide legend right hand side(in italics): "Waffenfabrik Walther Zella-Mehlis"
- 16 slide serrations

Special Markings:

- Misspellings in company address: e.g. SN 58588 inscribed with "Zel a-St Bl." instead of "Zella-St. Bl."

Use of Walther Pistols Mod. 7

The Walther pistol Mod. 7 became a popular pocket pistol because of its compact dimensions and its product qualities.

Officers and high-ranking military employees ("Militärbeamte") had to procure their service pistols privately. For them the Walther Mod. 7 was a popular defensive weapon, too.

Total production of Walther Mod. 7 pistols: about 45,000 pistols

Walther Mod. 8

Picture 10: *Walther Mod. 8, 1st variation, SN 404 293, caliber 6.35 mm Browning (.25 ACP)*

Production:	1920/21 - 1943 (?)
Caliber:	6.35 mm Browning = .25 ACP (8 rounds)
Total Length:	5.0 in.
Barrel Length:	2.8 in.
Weight:	12.7 oz.
Rifling:	6 grooves, right-twisted
Operation:	unlocked blowback action with shrouded hammer and firing pin

Preliminary remarks:

The Walther pistol Mod. 8 has some technical features, which are basically still the same as the ones of Mod. 2 to Mod. 7. Its outer design, however, already uses the design features of the later Walther PP.

The Walther Mod. 8 is the first Walther pistol to carry the model designation in its slide legend.

On 24th June 1919 Walther filed a patent protecting a "Handfeuerwaffe mit Hahnschloss" (hand gun with hammer breech action), as realized in Mod. 8. This patent was granted on 13th September 1920 with patent number 325 374.

Patent number 334 042 applied on 15th July 1919 only for a "Selbstladepistole mit feststehendem Lauf" (self-loading pistol with a fixed barrel), but in the text it becomes clear that this claim includes the pivoting trigger guard, which serves as a dismantling device. This patent was granted on 9th March 1921. To be on the safe side there was another patent (number 335 901) filed on 30th November 1919 to protect this pivoting trigger guard.

Mod. 8

Characteristics:

- Slide legend left hand side (in italics): "Walther's Patent Mod.8" with Walther banner
- Slide legend right hand side (in italics): "Waffenfabrik Walther Zella-Mehlis"
- Proof marks: "Crown over N", from January 1940: "Eagle over N"
- Serial number horizontally on right hand side of frame behind grip plate
- Fixed barrel, which serves as a guide for the recoil spring
- Pivoting trigger guard, which serves as a dismantling device
- Sights: sighting rib with milled front and rear sights
- 16 slide serrations on both sides of the slide
- Ejection port on right side of slide
- Shrouded hammer
- Internal trigger bar
- Bottom magazine release
- Long safety lever on left hand side of frame in front of the grip plate meshing with the sear –
- Cf. patent number 325 374 "Handfeuerwaffe" (handgun), filed on 24th June 1919, granted on 13th September 1920
- Patent number 390 298 (see above) claiming protection for "Handfeuerwaffe mit Hahnschloss" (hand gun with hammer breech action) includes a grip safety, too. But this was not realized in Mod. 8.
- Grip plates made of black, checkered, hard rubber; ivory or mother-of-pearl grips optional at an extra charge;
- Grip plates fitted to the frame with enamel medallions (left hand side with intertwined "CW", right hand side with "6,35" - patent number 319 461 filed on 24th June 1919 and granted on 8th March 1920, plus an additional patent filed on 13th December 1919)
- Magazine follower as a punched and bent sheet metal part, gap between end of follower and magazine body
- Some magazines with a capital "W" milled into the bottom
- Metal finish: salt blued (first time high-gloss polished finish as a standard!), nickel-plated finish or engravings optional at an extra charge
- The Walther literature mentions some rare Mod. 8s with lightweight frames.

Mod. 8, 1st Variation (1920 – 1926)

SN 390 000 to 472 000

Characteristics:

- Internal extractor (cf. patent number 313 011, filed on 25th May 1918, granted on 24th June 1919)
- Latch on right hand side of pivoting trigger guard (cf. patent number 334 042, filed on 15th July 1919, granted on 9th March 1921 and additional patent number 335 901 filed on 30th November 1919)
- Firing pin in a box-type housing (cf. Patent Number 334 448, filed on 15th July 1919, granted on 15th March 1921)
- Grip plates mostly without Walther banner at lower edge
- Grip plates fitted to the frame by enamel medallions (cf. patent number 319 461, filed on 24th June 1919, granted on 8th March 1920 and additional patent number 326 373, filed on 13th December 1919), left medallion with intertwined "CW", right one with "6,35"
- Some specimens lack the "S" resp. "F" at the safety lever.
- Magazine floor plate with "W" (in a circle), some (early?) specimens with a wide, engraved "W"
- About 84 300 Mod. 8s of the 1st variation were produced.

Mod. 8, 2nd Variation (1927 – 1933)

SN 465 000 to 486 450 (serial number range overlaps with 1st variation) and

700 000 to 714 400 (There is a gap in the serial number range due to a reorganization of the serial number system)

Characteristics: as before, but

- Exposed extractor
- Grip plates mostly with Walther banner at lower edge
- Magazine floor plate mostly with Walther banner instead of "W" (in circle)
- About 24 200 Mod. 8s of the 2nd variation were produced.

Mod. 8, 3rd Variation (1934 – 1943?)

SN 714 400 to 743 700 and 91 001A to 96 900A (from 1938 new serial number system with "A" for "Acht")

Characteristics: as before, but

- Pivoting trigger guard no longer locked by a latch but spring loaded (cf. later Walther PP and PPK)
- Firing pin in a U-shaped housing (cf. early Walther PP with two-piece firing pin)
- Proof marks: up to January 1940 "Crown over N", from January 1940 "Eagle over N" (from about SN 91 900A), however: up to SN 95 300A the proof mark "Crown over N" was still used sporadically
- SN 722 300 to 727 700: massive SN-block of Mod. 8s 2nd variation, i.e. with a latch at the trigger guard. The reasons for this are unkown.
- About 37 200 Mod. 8s of the 3rd variation were produced.

Special models resp. Markings:

- Stamp "Germany" or "Made in Germany" on pistols for export purposes
- Some Mod. 8 pistols have longer barrels e.g. for the export to Austria-Hungary to comply with gun laws there (e.g. SN 23 615).
- SN475 282: target pistol with longer grip, wooden grip plates and adjustable rear sight
- SN 700 050: additional inscription on the slide "Generalfeldmarschall Der Luftwaffe / H. Göring". Due to the spelling mistake the authenticity is questionable.
- SN 729 793: additional inscription on right hand side of the slide "Dr. Goebbels" and "Eagle over Swastika" (with additional letters "R" and "K") - authenticity is questionable
- SN 715 763: marked "St. P.V. D31" was issued to "Geheime Staatspolizei" ("Gestapo") in a serial-numbered belt holster
- SN 718 823: issued to a "Reichsforstbeamte" (forestry official) in a green pistol belt holster
- SN 566 361: used as a service pistol by Sachsen-Anhalt police (East Germany) – even after 1945.

Use of Walther Mod. 8 Pistols:

The Walther pistol Mod. 8 became a popular pocket pistol resp. side arm (carried in a belt holster) because of its compact dimensions, its magazine capacity and its superb quality respectively finish.

Officers and high-ranking military employees ("Militärbeamte") had to procure their service pistols privately. For them the Walther Mod. 8 was a popular side arm, too.

Often this model was carried as a concealed "backup" weapon. After his imprisonment in Landsberg/Bavaria in 1923 Adolf Hitler seems to have carried such a Mod. 8 in a concealed pocket in the inside of his trouser leg, too. Allegedly his niece, Angela ("Geli") Raubal, committed suicide with this gun in 1931.

Many police forces purchased Mod. 8s as service pistols – esp. for plain clothes police.

The "Geheime Staatspolizei" ("Gestapo") and "Reichsforstbeamte" (forestry officials) used Mod. 8s as well – see above.

End of Mod. 8 production:

Presumably there were no new Mod. 8 pistols produced from about 1940 to 1943 (?). In these years there were most probably such pistols assembled from parts only, which had been produced before.

Total production of Walther Mod. 8 pistols: about 145,700 pistols

Copies of Walther Mod. 8 pistols:

Although the Walther Mod. 8 was a very popular pistol, there were no copies produced by other manufacturers – in contrast to Mod. 9.

Walther Mod. 9

Production:	1921 - 1943 (?)
Caliber:	6.35 mm Browning = .25 ACP (8 rounds)
Total Length:	4.0 in.
Barrel Length:	2.0 in.
Weight:	9.0 oz.
Rifling:	6 grooves, right-twisted
Operation:	hammerless, unlocked blowback action, striker-fired - similar to Mod. 2

Preliminary remarks:

The Walther pistol Mod. 9 has some technical features, which are basically the same as the ones of Mod. 1 – e.g. the slide, which is open at the top and the hammerless, striker-fired action.

Like the Mod. 8 the Mod. 9 carries the model designation in its slide inscription. The internal company designation for the variation without the etched engraving was Mod. 9a, whereas the one with this etching was designated as Mod. 9b. These designations cannot be found in the slide inscription, however.

Mod. 9a

Characteristics:

- Slide legend left hand side (in italics): "Walther's Patent Mod.9" with Walther banner
- Slide legend right hand side (in italics): "Waffenfabrik Walther Zella-Mehlis (Thür.)"
- Proof marks: "Crown over N", from January 1940: "Eagle over N"
- Serial number: first horizontally on right hand side of frame behind grip plate, later vertically in front of left gip plate (next to the safety lever)
- Stepped slide, which embraces the barrel sleeve only partially and is open at the top
- Fixed barrel, pressed into the barrel sleeve, which is part of the frame and reaches to the front of the frame
- Recoil spring is held in a separate recess below the barrel
- Sights: sighting groove on top of the slide and milled front sight on top of the barrel sleeve
- 17 or 18 slide serrations on both sides of the slide
- Slide dismantling lever integrated in 8-shaped slide catch at the rear of the slide
- Ejection to the top
- Extractor first internal, later exposed
- Signal pin (not loaded chamber indicator!) protrudes from 8-shaped slide catch, when pistol is cocked

Picture 11: *Walther Mod. 9, 1st variation, SN 416 119, caliber 6.35 mm Browning (.25 ACP)*

- Internal trigger bar
- Bottom magazine release
- Long safety lever on left hand side of frame in front of the grip plate meshing with the sear –
- Cf. patent number 325 374 "Handfeuerwaffe" (handgun), filed on 24th June 1919, granted on 13th September 1920 – cf. Mod. 8
- Patent number 390 298 (see above) claiming protection for "Handfeuerwaffe mit Hahnschloss" (hand gun with hammer breech action) includes a grip safety, too. But this was not realized in Mod. 9.
- Late Mod. 9s mostly with magazine safety, which meshes with the trigger bar
- Grip plates made of black, checkered, hard rubber; ivory or mother-of-pearl grips optional at an extra charge
- Metal finish: salt blued (high-gloss polished finish), nickel-plated finish or engravings optional at an extra charge

Mod. 9, 1st Variation

SN 410 001 to 428 000

Characteristics:

- Internal extractor (cf. patent number 313 011, filed on 25th May 1918, granted on 24th June 1919)
- Trigger spring as a spiral spring
- Recoil spring held in front of slide by a mushroom-shaped, separate cone
- No magazine safety
- Proof marks: "Crown over N"
- Serial number on the right hand side of the frame (behind the grip plate) and the last three digits of the serial number stamped onto the underside of the slide (in the muzzle area)
- Sights: narrow sighting groove on top of the slide and milled front sight on top of the barrel sleeve
- Some early specimens in the SN range 41x xxx have a square cutout in the front

of the magazine well or no cutout at all – instead of the rounded cutout for the removal of the magazine.

- Grip plates mostly without Walther banner at lower edge
- Grip plates fitted to the frame by enamel medallions (cf. patent number 319 461, filed on 24th June 1919, granted on 8th March 1920 and additional patent number 326 373, filed on 13th December 1919), left medallion with intertwined "CW", right one with "6,35"
- Magazine follower mostly as a deep-drawn sheet metal part
- Magazine floor plate without Walther banner
- Some Mod. 9 pistols without the "S" and "F" markings near the safety lever.

Mod. 9, 2nd Variation

SN 428 000 to 596 900

Characteristics: as before, but

- Exposed extractor
- Sights: from about SN 503 000 wide sighting groove on top of the slide and milled front sight on top of the barrel sleeve
- Grip plates mostly with Walther banner at lower edge
- Magazine follower as a punched and bent sheet metal part, gap between end of follower and magazine body
- Magazine floor plate: from about SN 438 000 mostly with Walther banner

Mod. 9, 3rd Variation

SN 530 000 to 615 000 (serial number range overlaps with 2nd variation)

Characteristics: as before, but

- Trigger spring as a triangular-shaped torsion spring
- Grip plates mostly screwed to the frame with blued metal discs, some specimens still with enamel medallions
- Some specimens with magazine safety
- Walther banner mostly at lower edge of grip plates and on magazine bottom plate

Mod. 9, 4th Variation

SN 615 000 to 652 000 and 192 001N to 197 960N (from 1938 new serial number system with "N" for "Neun")

Characteristics: as before, but

- Serial number vertically in front of left gip plate (next to the safety lever)
- Grip plates mostly screwed to the frame with blued metal discs, some specimens still with enamel medallions
- Late specimens already with magazine safety
- Recoil spring no longer held in front of slide by a mushroom-shaped, separate cone, but front end of spring sits on a pin in the front of the slide
- Magazine follower as a punched sheet metal part, but end of follower is folded into the magazine body
- From SN 650 000 most specimens with magazine safety
- Walther banner mostly at lower edge of grip plates and on magazine bottom plate
- Proof marks: up to January 1940 "Crown over N", from January 1940 "Eagle over N" (from about SN 195 000N), however: up to SN 197 600N the proof mark "Crown over N" was still used sporadically
- Some late specimens of Mod. 9 produced by assembly of parts, which had been produced long before (variations 1 to 3)

Mod. 9b (no separate serial number range)

SN 502 045 to 650 530 and 196 105N to 196 950N

Characteristics: as before, but

- Etched engraving on both sides of the slide

Special models resp. Markings:

- Stamp "Germany" or "Made in Germany" or "Allemagne" on specimens for export purposes
- Some Mod. 9 pistols have longer barrels e.g. for the export to Austria-Hungary to comply with gun laws there.
- SN xx5 643 (the first two digits are milled out): inscribed on left hand side of slide with: "Die denkende Schusswaffe" (The intelligent firearm); double action, which permits carrying the pistol loaded, but uncocked (Walther/Ulm company museum)
- SN 410 530: marked on slide "OPM 274" (Norwegian police, Oslo Police Museum)
- SN 414 144: nickel-plated finish, formerly owned by Carl Walther, captured on 6th April 1945 in Zella-Mehlis by Captain Clemens A. Werner
- SN 535 724: gold finish, ivory grip plates, left grip plate with inlaid silver badge saying "Gewidmet von der Firma Carl Walther Zella-Mehlis 4.9.1924" (dedicated by Carl Walther ...)
- SN 544 892: engraved (scrolls), blued finish, rubber grip plates, left grip plate with inlaid silver badge saying "Erinnerung an Tontaubenschießen vom 7.9.1924 v. Bürgerl. Jägerver. Würzburg gestiftet von Waffenfabrik Walther" (shooting prize of hunters' association Würzburg for clay pigeon shooting)
- SN 570 894: marked "NSFK" ("Nationalsozialistisches Fliegerkorps" – authenticity is questionable
- SN 609 997: marked "Bayern Stadtpolizei Bamberg" (on left hand side of slide) – pistol was reused by Bavarian municipal police in Bamberg after 1945
- SN 641 997: engraved (scrolls), nickel-plated finish, wolf's head inlaid in engraving – allegedly formerly owned by Eva Braun (Adolf Hitler's spouse)
- SN 647 565: engraved (oak leaves), blued finish, initials "AH" on backstrap – allegedly formerly owned by Adolf Hitler
- SN 566 361 and 194 462N: used by Sachsen-Anhalt police (East Germany) after 1945
- Several Mod. 9s marked on top of slide: "Chas. Heyer&Co., Nairobi/Kenia" – sold to an African hunters' shop (hopefully not for hunting elephants!)

Use of Walther Mod. 9 Pistols:

The Walther pistol Mod. 9 became a popular pocket pistol resp. side arm (carried in a holster) because of its compact dimensions and its superb quality resp. finish.

Without any protruding parts and with its compact dimensions the Mod. 9 was designed as a vest-pocket or a "ladies'" pistol. There was even a purse-type little bag produced to accommodate the Mod. 9 in any lady's handbag.

Furthermore the Mod. 9 was a popular "backup" weapon for anyone who carried a service pistol on duty.

Several police forces in West and East Germany used the Mod. 9 as a service pistol – before and after 1945 (esp. for plain clothes police).

Often this model was carried as a concealed "backup" weapon, too, next to a service pistol in a bigger caliber.

End of Mod. 9 production:

Presumably there were no new Mod. 9 pistols produced from about 1940 to 1943 (?). In these years there were most probably such pistols assembled from parts only, which had been produced before.

Total production of Walther Mod. 9 pistols: about 248,000 pistols

Copies of Walther Mod. 9 pistols:

As the Walther Mod. 9 was a very popular pistol, there were quite a few, unlicensed copies produced by other manufacturers – e.g.:

Bernadelli Mod. VP ("Vest Pocket") – Vincenzo Bernadelli, Gardone/Italy (from1949)

This "Mod. VP" has the following characteristics in common with the Walther Mod. 9:

- Contours of frame and slide
- Caliber 6.35 mm Browning (.25 ACP)
- Stepped slide, which embraces the barrel sleeve only partially and is open at the top
- Fixed barrel, pressed into the barrel sleeve, which is part of the frame and reaches to the front of the frame
- Similar trigger mechanism
- Recoil spring under the barrel
- Bottom magazine release
- 8-shaped slide catch at the rear of the slide
- Long safety lever on left hand side of frame in front of the grip plate

Bernadelli Mod. Baby – Vincenzo Bernadelli, Gardone/Italy (from1949)

This "Mod. Baby" is basically a downsized "Mod. VP", with the same characteristics, but in caliber .22 l.r.

Bernadelli Mod. 69 – Vincenzo Bernadelli, Gardone/Italy (from1969)

This "Mod. 69" is a modernized "Mod. VP" with the same basic characteristics, but in the calibers .25, .22 l.r. and .22 short.

Reck P8 – Karl Arndt Reck, Lauf near Nuremberg (from1955)

This model "Reck P8" has the following characteristics in common with the Walther Mod. 9:

- Contours of frame and slide
- Caliber .25 ACP
- Stepped slide, which embraces the barrel sleeve only partially and is open at the top
- Fixed barrel, pressed into the barrel sleeve, which is part of the frame and reaches to the front of the frame
- Similar trigger mechanism
- Recoil spring under the barrel
- Bottom magazine release
- 8-shaped slide catch at the rear of the slide

In the USA this "Mod. P8" was sold as "LA Fury" or "Chicago Club". Today the right to the name "Reck" is owned by Umarex/ Arnsberg – a subsidiary of Walther/Ulm. Thus things have come full circle.

Model "Perla" – Frantisek Dusek, Opocno/Czech Republic

Model "Perla" was made from 1935 to 1941 in .25 ACP and is a "lookalike" of the Walther Mod. 9.

Model "Liliput" – August Menz, Suhl/ Thuringia

Model "Liliput" was made from 1920 onwards in the calibers 4.25 mm and 6.35 mm (.25 ACP). In 1937 the company Lignose/ Berlin bought Menz and carried on with their pistol production.

The "Liliput" is only a Walther Mod. 9 copy in the wider sense, as it differs technically quite a lot.

Table: Serial Number Ranges Walther Mod. 1 through 9

Model	Pistols Produced	Serial Number Ranges	Years of Production
Mod. 1	about 30 000	1 – 30 800	1910 – 1915 (?)
Mod. 2	about 19 000	1 – 72 000	1913 – 1915
Mod. 3	about 3 500	11 000 – 34 100	1913 – 1915 (?)
Mod. 4	about 351 000	20 000 – 72 000 72 001 – 289 000 432 000 – 493 700	1913 – 1929 (?)
Mod. 5	about 120 000	26 000 – 102 000 368 100 – 388 700	1915 – 1917
Mod. 6	about 1000	1 – 1 038	1915 – 1917
Mod. 7	about 57 500	4 000 – 5 410 27 800 – 83 900	1915 – 1919
Mod. 8	about 145 700	700 000 – 743 700 91 001A – 97 000A	1920 – 1943 (?)
Mod. 9	about 195 000	410 000 – 651 800 192 000N – 198 000N	1921 – 1943 (?)

Remarks:

- Mod. 1 pistols had a separate serial number range.
- Mod. 2 and Mod. 5 pistols shared the same serial number range up to SN 72000 (1915?).
- Total production of Mod. 1 to Mod. 9 pistols: about 975 000 pistols
- (Note: Not for every serial number, which was issued, there was a pistol delivered, as some did not pass the final quality control or proof).

Walther PP and PPK Zella-Mehlis Preliminary Remarks

It is difficult to keep the development of these two models apart, as many technical changes of the Walther PP were taken over in the production of the Walther PPK. So later in the text the term "PP/PPK" will precede any information referring to both models. Otherwise it will be "PP" or "PPK" only.

Extremely rare models like the "KPK" or the "Ultra" will not be dealt with, as they are beyond the financial means of most collectors. Furthermore any information on these models is simply too vague.

Walther PP/PPK Production Process:

The main parts were made by Walther/ Zella-Mehlis. Minor parts were mostly produced by Thuringian subcontractors – e.g. ejectors provided by Oskar Weihrauch in Zella-Mehlis.

The assembly of PPs/PPKs was done in the so-called "Schwarzmontage" ("black" assembly"), i.e. assembly of blued parts in piecework. In this assembly shop workers sat in four rows and assembled pistols for 40 Pfennige per piece. All the major and minor parts were delivered to the assembly lines in boxes containing all the parts for 50 pistols each. The assembled pistols then passed two quality control checks.

Walther PP/PPK Serial Numbers and Proof-firing:

It is extremely difficult to date PPs/PPKs according to the serial numbers – for several reasons:

- Walther did not always produce a pistol for each consecutive serial number.
- During the production process whole batches of pistols with consecutive serial numbers were set aside for future use – e.g. engravings or special contracts. Then the dates of production and proof firing can differ. This means that any PP/PPK with a serial number indicating a production before January 1940 may nonetheless have the "Eagle over N" proof marks.
- The proof law, which determined the transition from "Crown over N" to "Eagle over N" was enacted halfway through the year 1939, but implemented as late as January 1940 only.

Until the introduction of the suffixes "P" and "K" both models were consecutively numbered in a joint serial number range.

Some serial number blocks, however, were used exclusively for special contracts – e.g. for the PPKs of the "Reichszeugmeisterei" ("RZM") and the PPs resp. PPKs of the "Polizeidirektion München" ("PDM") – see there.

In 1938 (presumably in April) Walther had to change their serial number system for two reasons:

- The "magical" number 1 000 000 was almost reached.
- Because of the arming of the German "Wehrmacht" (army, navy and air force) Walther became a large armament concern and had to restructure their operating system and their forms of organization.

The suffix "P" was added to the serial numbers of the Walther PP, resp. "K" to the ones of the Walther PPK. This alteration was not carried out at once and consistently though:

Walther PP: In the serial number range from 997 950 to about 998 080 there was simply the suffix "P" added to the serial number. Then it was left away again – e.g.

SN 998 940 to 998 960. From about 998 784P to about 999 930P the first digit was omitted and the serial number had only five digits left – e.g. [9]99 940P. From 100 000P to about 399 000P the numbering was carried out consistently up to the last days of World War Two, when the suffix "P" was dropped again. The very last PPs, which were ever made, allegedly had the suffix "R" instead of "P". But there is no proof or piece of evidence for that.

Walther PPK: Beyond serial number 999 800 there were about 9 670 PPKs produced with seven digit numbers (1 000 001 to 1 009 670) – the so-called "million series". From serial number 172 000K onwards the numbering was carried out consistently up to 431 097K – with one exception: In 1939 there were two serial number blocks (about 4800 PPKs) where the suffix "K" was replaced by "W" (SN 185 440W – 189 270W and SN 193 930W – 195 710W). The reasons for this are unknown. The assumption that "W" stands for "Wehrmacht" needs proof.

The very last PPK which is known carries the suffix "A": SN 433 521A.

In the serial number range 191 160K to 191 200K there were about 40 PPKs produced, which erroneously were marked "Mod. PP" instead of "Mod. PPK" on the slide.

After 1945 Walther PPs were assembled in the East German factory "VEB Ernst Thälmann"/Suhl using pre-numbered parts, which were found in old stocks, with serial numbers from about 397 000 to 399 990.

Matching Walther PP/PPK Serial Numbers:

The matching of frame and slide is not only important because of originality, but also for another reason: Before the blueing process the Walther PPs and PPKs were assembled in a half-finished condition, then grinded and polished to make sure that especially the area at the rear of slide and frame was smooth and without a step.

In the production period from 1929 until 1941/42, before the serial number was applied additionally to the right hand side of the slide, the Walther workers had a problem after the blueing process. They had to match the slides and frames again. So they used a trick: After the polishing process they scratched the last three digits of the serial number into the interior of the slide opposite the ejection port. This was done by hand and with a sharp needle. These three digits are hardly legible, as they were scratched in only lightly and then blued over.

Total Walther PP/PPK Production

From 1929 to early 1945 a total of about 830 900 PPs and PPKs were produced. An estimated 65 % of that number were PPs (i.e. about 540 000 PPs) and an estimated 35 % were PPKs (i.e. about 290 900 PPKs). These figures can only be estimates, as all the Walther company files are presumably either lost or even destroyed.

The matching of serial numbers and year of production therefore can only be made empirically and by interpolation (based on invoices, engraved dates, proof marks etc.).

The statements in this book are mostly based on the author's database, which contains the characteristics of about 7 200 PPs and 8 500 PPKs. The fact that there are more PPKs than PPs in this database, although more PPs than PPKs were produced, can be explained by the circumstance that the PPK perhaps found more interest in the collector's world – especially in the Anglo-American one.

The subdivision into different variations cannot be very sharp and distinctive, as most transitions are flowing – cf. Walther Mod. 1 to 9, too.

Walther PP/PPK Production - Beginning and End:

The first presentation of a Walther PP prototype was in June 1925 at the "Internationale Polizeitechnische Ausstellung" (a police equipment fair) in Karlsruhe/

Germany. It is unknown whether this was already a working gun or not. The first design drawings are dated December 1928.

There seems to be a common consent only as to the beginning of the production of Walther PPs in 1929 with serial number 750 001, and of Walther PPKs in 1930/31 with serial number 755 000. The actual series production of PPKs began with serial number 760 000 only.

The end of the production of Walther PPs seems certain: April 1945 with roughly serial number 397 000P, when the Walther factory was occupied by US-American troops. The very last PPs were assembled in the factory for US-American soldiers (so-called "US assemblies" or "cigarette pistols"). Those pistols mostly have no matching numbers and no proof marks and some of them are still in-the-white even.

The end of the production of Walther PPKs is less certain. Some (above all US-American) authors tend to believe that the actual series production ended in 1943/44 already. They conclude this from the following evidence:

- Whereas the quality of Walther PPs was gradually deteriorating from 1943 onwards and cheaper materials were used (thinner blueing, grip plates made of pressed wood, rougher machining, etc.), such signs of "degeneration" can hardly be found with the Model PPK pistols – perhaps because PPKs were no longer produced but only assembled from stored parts towards the end of World War Two.
- In 1944 the company address on the left side of the slide of PPKs was not omitted and replaced by the "ac" code on the right side of the slide – as it was the case with Walther PPs.
- The Walther PP was "fitter for duty" as a service pistol due to the bigger magazine capacity, the longer barrel length and the less fragile, two-piece grip plates.
- By the end of World War Two the Model KPK was perhaps intended to replace the PPK. The KPK was of a simpler design and had two-piece grip plates.

The last PPKs to leave the factory in 1945 were obviously no products of the series production, but hastily assembled guns made of parts from old stocks. From serial number 433 500 onwards there seem to have been PPKs with the suffix "A" to the serial number. But there is only one piece of evidence known: SN 433 521A – if it is authentic.

Towards the end of World War Two Walther strangely enough produced a series of PPKs with lightweight ("Dural") frames and slides with a high-gloss polished finish (SN range: 417 150K – 417 385K) and later even a series of PPKs in caliber .22 l.r., which were "unfit for service" (SN range: 430 600K – 430 960K) as well as PPKs in caliber 7.65 mm Browning (.32 ACP) mostly with "Dural" frames and some of them even with chrome- or nickel-plated finish – sometimes still without proof marks (SN range: 430 969K – 431 097K).

There are two interesting, but unproven explanations:

- High quality parts, which were kept in storage, were used to assemble these PPKs.
- Walther made provisions for a prospective commercial market after the end of World War Two.

The latter theory sounds perhaps a bit daring, but it is backed by the fact that by the end of World War Two Walther made the effort of producing a costly color brochure (!) for this variation.

The last known and reported PPK made in Zella-Mehlis is the one with serial number 433 521A.

Walther PP/PPK Grip Plates, Magazines and Magazine Extensions

It is very difficult to classify grip plates and magazine extensions on an empirical basis, as these parts were presumably often exchanged and interchanged. There are hardly any reliable data. Therefore the following statements can only be an attempt at a classification and a temporal assignment.

The producer of PP/PPK grip plates was most probably a subsidiary of Walther/Zella-Mehlis: a former paper mill in Wasungen near Meiningen/Thuringia – at least from the mid-1930s onwards.

When the new "Reichsbetriebsnummern" (RBN) were issued, on 25th June 1943 Walther was given the number RBN 0/0345/0044, whereas the subsidiary in Wasungen was given the number RBN 0/0347/0056.

Walther PP grip plates

- **1st variation** (1929 – 1933?): made of black, hard rubber with a thick, X-shaped reinforcement inside; outside checkered, with Walther banner
- **2nd variation** (1931 – 1934?): made of black, hard rubber with only a thin, X-shaped reinforcement inside; outside checkered, with Walther banner
- **3rd variation** (1934? – 1938?): made of black hard rubber with cast-in metal reinforcements (in the US-literature often named "weighted grips"; this designation is simply wrong, as these metal plates are nothing but reinforcements); outside checkered, with Walther banner
- **4th variation** (1938? – 1944): made of black or reddish brown "Trolitan" with a cavity inside; outside checkered, with Walther banner ("Trolitan" is a molded plastic based on phenolic resin – similar to "Bakelite"). According to Walther design drawings those grip plates should have been used from 1st September 1938 onwards already.
- **5th variation** (1944 – 1945): made of pressed beech wood, checkered with Walther banner.

Remarks:

Compared to the grip plates used by Walther/Ulm the ones used by Walther/Zella-Mehlis show a larger Walther banner.

White grips or those made of special material (ivory, mother-of-pearl, carved wood etc.) were mostly used with engraved PPs at an extra charge.

Reddish brown grips were destined for target shooting teams of the SA. Each member of an SA shooting team was given one pair of reddish brown grips and each year an especially selected PP by Walther. The reddish brown grips were then changed over to the new PP each year.

Grips made of mottled brown "Trolit" (cf. PPK grips), not made of "Trolitan", were mostly used with engraved PPs – even with PPs of the first variation.

In the early 1930s grips of the first or the second variation were obviously used alternatively.

Walther PPK grips

- **1st variation** (1931 – 1932): one-piece wrap-around type grips made of mottled brown "Trolit" ("Trolit" is an injection-molded plastic based on cellulose acetate) without a cast-in metal reinforcement in the lower backstrap; outside checkered, with Walther banner; grip screw from the left to the right;
- **2nd variation** (1932 – 1935): as before, but with a cast-in metal reinforcement in the lower backstrap
- **3rd variation** (1935 – 1940): as before, but with grip screw from the right to the left
- **4th variation** (1940 – 1945): as before, but color of plastic material changing from mottled brown to chocolate brown, then grey-brown and finally black

Remarks:

The grips of all PPKs produced by Walther/Zella-Mehlis are all of the one-piece wrap-around type in contrast to the two-piece grips used by Walther/Ulm later. Very early Manurhin PPKs, however, still show honey-colored to dark brown one-piece wrap-around grips.

White grips or those made of special material (ivory, carved wood etc.) were mostly used with engraved PPKs at an extra charge.

Walther PP/PPK Magazines

Magazines were accidentally interchanged even more often than grips, which today leads to great problems when trying to allocate a magazine to a pistol or to assign a magazine to a year of production.

- Walther PP (7.65 mm Browning - .32 ACP) 1st variation: matte finish; magazine body with Walther banner only on left hand side; magazine follower as a punched and bent sheet metal part, gap between end of follower and magazine body (cf. Walther Mod. 8s and early Mod. 9s); seven viewing holes on both sides; crescent shaped cutout for the magazine release at the upper front of the body (not true for the "Polizeidirektion München" ("PDM") PPs due to their bottom magazine release – see there)
- Walther PP (7.65 mm Browning - .32 ACP) 2nd variation: as before, but polished finish; magazine follower as a punched and deep-drawn sheet metal part, no more gap between end of follower and magazine body, as the follower is folded into the magazine body; seven viewing holes on both sides
- Walther PP (7.65 mm Browning - .32 ACP) 3rd variation: as before, but high-gloss polished finish
- Walther PP (7.65 mm Browning - .32 ACP) 4th variation: as before, but Walther banner on left hand side of magazine body with additional model and caliber designation ("PP 7,65 m/m" – spelling with a slash)
- Walther PP (7.65 mm Browning - .32 ACP) 5th variation: as before, but dull, military finish
- Walther PP (7.65 mm Browning - .32 ACP) 6th variation: as before, but stamp "w" on right hand side of magazine body ("w" for "weich" – i.e. magazine lips not hardened); from 11th June 1942 according to the design drawing, but obviously not implemented before 1944
- Walther PP (9 mm kurz - .380 ACP) 1st variation: as 1st variation cal. 7.65 mm magazines, but matte finish; magazine body with Walther banner only on left hand side; magazine follower as a punched and deep-drawn sheet metal part, no gap between end of follower and magazine body, as the follower is folded into the magazine body; crescent shaped cutout for the magazine release at the upper front of the body due to the magazine release push button at the upper left hand side of the frame (cf. PP in caliber 7.65 mm); seven viewing holes on both sides; some specimens with caliber designation "Cal. 9m/m" subsequently engraved on the back of the magazine
- Walther PP (9 mm kurz - .380 ACP) 2nd variation: as before, but some specimens with polished finish; no crescent shaped cutout for the magazine release at the upper front of the body due to the bottom magazine release; magazine follower mostly as a milled and no longer as a punched sheet metal part (from November 1940?); six viewing holes on right hand side, one hole only on left hand side
- Walther PP (9 mm kurz - .380 ACP) 3rd variation: as before, but high-gloss polished finish; magazine follower as a milled and no longer as a punched sheet metal part
- Walther PP (9 mm kurz - .380 ACP) 4th

variation: as before, but Walther banner on left hand side of magazine body with additional caliber designation ("Cal. 9 m/m" – spelling with a slash; rarely "Cal. .380"); some specimens with caliber designation "7,65 m/m" overstamped with "9 m/m" or ".380"; either seven viewing holes on both sides or six viewing holes on right hand side and one hole only on left hand side

- Walther PP (9 mm kurz - .380 ACP) 5th variation: as before, but dull, military finish
- Walther PPK (7.65 mm Browning - .32 ACP) 1st variation: matte finish; magazine body with Walther banner only on left hand side; magazine follower as a punched and deep-drawn sheet metal part, no gap between end of follower and magazine body, as the follower is folded into the magazine body; six or seven viewing holes on both sides; crescent shaped cutout for the magazine release at the upper front of the body (not true for the "Polizeidirektion München" ("PDM") PPKs due to the bottom magazine release – see there)
- Walther PPK (7.65 mm Browning - .32 ACP) 2nd variation: as before, but mostly polished finish; mostly six viewing holes on both sides
- Walther PPK (7.65 mm Browning - .32 ACP) 3rd variation: as before, but high-gloss polished finish, Walther banner on left hand side of magazine body with additional model and caliber designation ("PPK 7,65 m/m" – spelling with a slash!); six viewing holes on both sides
- Walther PPK (7.65 mm Browning - .32 ACP) 4th variation: as before, but dull, military finish
- Walther PPK (9 mm kurz - .380 ACP) 1st variation: matte finish; magazine body with Walther banner only on left hand side; crescent shaped cutout for the magazine release at the upper front of the body due to the magazine release push button at the upper left hand side of the frame (cf. PPK in caliber 7.65 mm); six viewing holes on both sides; magazine follower as a punched and deep-drawn sheet metal part
- Walther PPK (9 mm kurz - .380 ACP) 2nd variation: as before, but no crescent shaped cutout for the magazine release at the upper front of the body due to the bottom magazine release; six viewing holes on both sides
- Walther PPK (9 mm kurz - .380 ACP) 3rd variation: as before, but some specimens with polished finish; magazine follower mostly as a milled and no longer as a punched sheet metal part (from November 1940?); six viewing holes on right hand side, one hole only on left hand side
- Walther PPK (9 mm kurz - .380 ACP) 4th variation: as before, but high-gloss polished finish
- Walther PPK (9 mm kurz - .380 ACP) 5th variation: as before, but Walther banner on left hand side of magazine body with additional caliber designation ("Cal. 9 m/m" – spelling with a slash; rarely "Cal. .380"), some specimens with caliber designation "7,65 m/m" overstamped with "9 m/m" or ".380"; six viewing holes on right hand side and one hole only on left hand side
- Walther PPK (9 mm kurz - .380 ACP) 6th variation: as before, but dull, military finish
- Walther PP/PPK (.22 l.r.) 1st variation: neither Walther banner nor model or caliber designation on magazine body; crescent shaped cutout for the magazine release at the upper front of the body due to the magazine release push button at the upper left hand side of the frame (cf. PP/PPK in caliber 7.65 mm); magazine follower mostly as a milled and not as a punched sheet metal part; coil spring, seven viewing holes on right hand side and one hole only on left hand side
- Walther PP/PPK (.22 l.r.) 2nd variation: as before, but magazine body with a viewing slot (instead of viewing holes) on right

hand side, one hole only on left hand side, mostly milled follower, coil spring

- Walther PP/PPK (.22 l.r.) 3rd variation: as before, but pressed sheet metal follower, zig-zag spring
- Walther PP/PPK (6.35 mm Browning - .25 ACP): Mod. 8 magazine inserted into a PP/PPK magazine body, which serves as a "dummy" magazine body only

Remarks:

The shape of the Walther banner on PP/PPK magazines was subject to some changes from 1929 to 1945: from a big and flat shape to a smaller and more curved shape.

The size of the letters and digits in the model and caliber designations was subject to changes as well.

The single hole on the left hand side of the body of .22 and 9 mm kurz magazines is not a viewing hole. This hole was used during production as a reference hole for positioning the magazine body precisely in the machine.

Walther PP/PPK Magazine Extensions

Quite like whole magazines, magazine extensions were very often accidentally interchanged, which today leads to great problems when trying to allocate any magazine extension to a magazine and/or a pistol.

Basically PP and PPK magazine extensions differ only in the colors, which are normally the same as the ones of the grips plates. Exceptions here are the green and red extensions, which can be found with some magazines for caliber .22 l.r. (green) and 9 mm kurz (red). Allegedly extensions made of fluorescent plastic were produced as well for shooting at night.

Very early extensions were still made of metal, as can be seen in early Walther brochures. Later these extensions were made of plastic ("Trolit"). The so-called "boxed magazine extensions" were made of the same material – see below.

Walther delivered the PP/PPK pistols in a cardboard box with a sticker with matching numbers and with accessories (a small box for cleaning items, an oiler and dummy cartridges). One magazine had an extension, the spare one came without one. This is not true for PPs/PPKs for "Polizeidirektion München" ("PDM"). Their service pistols came with two magazines both with extensions. The PPs/PPKs for the "Wehrmacht" came with two magazines without extensions, however.

The shape of the extensions was subject to several changes:

- **1st variation**: flute at the bottom, smooth sides
- **2nd variation**: flute at the bottom, cavity on both sides,
- **3rd variation**: flute at the bottom, cavity on both sides, cutout at the rear for the bottom magazine release (for PPs/PPKs with top magazine release, too)
- **4th variation**: as before, but the front is drawn further down with a thin rim; hence the bottom no longer shows a flute, but a cavity
- **5th variation**: as before, but the front is drawn even further down with a thicker rim – similar to the ones for post-1945 magazines used by Walther/Ulm

For target shooting PP/PPK magazine extensions were provided, which were much longer and voluminous – the so-called "boxed magazine extensions". The ones for the PP and the ones for the PPK look similar, but cannot be interchanged.

Walther PP and PPK Prototypes

As mentioned above Walther first presented a Walther PP prototype in June 1925 at the "Internationale Polizeitechnische Ausstellung" (a police equipment fair) in Karlsruhe/Germany. It is not known whether this was already a working gun or not.

On 24th June 1919 Walther applied for a patent protecting a "Handfeuerwaffe mit Hahnschloss" (hand gun with a hammer breech action). This patent was granted on 13th September 1920 with patent number 325 374.

A patent for a two-piece firing pin like the one used in early PPs was already filed on 24th August 1919. This patent with number 323 398 was granted on 24th July 1920.

On 15th July 1919 Walther filed patent number 334 042 for a "Selbstladepistole mit feststehendem Lauf" (self-loading pistol with a fixed barrel), but in the text it becomes clear that this claim includes the pivoting trigger guard, which serves as a dismantling device. This patent was granted on 9th March 1921. To be on the safe side there was another patent (number 335 901) filed on 30th November 1919 to protect this pivoting trigger guard.

On 28th March 1923 Georg Walther filed a patent protecting a "self-loading pistol with hammer breech action". This patent was granted with number 390 298 on 16th February 1924.

An additional patent number 520 437 was granted on 11th March 1931 for a "hammer breech action for self-loading pistols" (with a de-cocking function of the safety lever).

On 22nd January 1924 Fritz Walther filed a patent claiming protection for a "self-loading pistol with magazine and double-action trigger". This patent with number 446 763 was granted as late as 11th July 1927. On 7th June 1925 Walther applied for an additional patent claiming protection for "double-action and a pivoting safety lever in the slide". This patent with number 433 937 was granted on 15th September 1926. A similar patent claiming protection for a "pivoting safety lever with de-cocking function" was already filed on 10th April 1927, but was granted only on 11th March 1931.

An additional patent application to protect a "firing pin, which covers and blocks the hammer" was filed on 7th November 1939, but granted only on 16th June 1933.

On 20th November 1933 Walther was granted patent number 586 322 protecting a "self-loading weapon with interchangeable barrels" (similar to the later Heckler & Koch Mod. HK4 from 1967). But this novelty was not realized in the production of PP/PPK pistols.

Walther kept a close watch over the safeguarding of their own interests and patent rights. So e.g. they forbade the company Menz in Suhl by a court decision dated 13th April 1938 to call their 1936 pistol model "Pistole PB" because of the similarity of this name to their "Pistole PP" – although "PB" here should stand for "Polizei und Behörde" (police and other authorities).

We can assume that several designers contributed to the development of the Walther PP – perhaps even simultaneously. So the following consecutive numbering system of Walther PP prototypes does not necessarily mean a chronological sequence. This listing does not claim completeness either.

Walther PP Prototype 1

Characteristics:

- Slide legend left hand side (in italics): "Walther's Patent Cal. 7,65" with Walther banner below – similar to Mod. 8
- Slide legend right hand side (in italics): "Waffenfabrik Walther Zella-Mehlis (Thür.)" – similar to Mod. 8

- No serial or project number
- Pivoting trigger guard with locking latch – similar to Mod. 8
- Unlocked blowback action with shrouded hammer and firing pin – similar to Mod. 8
- Internal extractor – similar to Mod. 8 1st variation
- Grip plates fitted to the frame by enamel medallions (cf. patent number 319 461, filed on 24th June 1919, granted on 8th March 1920 and additional patent number 326 373, filed on 13th December 1919 – similar to Mod. 8
- All in all the contours and design characteristics remind the viewer of an enlarged Walther Mod. 8 or even an Ortgies pistol.

This prototype 1 shows the following novelties, however:

- Magazine release as a push button on the upper left hand side of frame
- Pivoting safety lever on left hand side of frame (behind the grip plate) – similar to Mod. 2 to Mod. 7, but contra-rotating
- Grip safety (Patent number 390 298 claiming protection for "Handfeuerwaffe mit Hahnschloss" (hand gun with hammer breech action) includes a grip safety, too. But this was not realized in any Walther pistol so far.)
- "automatic" (i.e. shooter independent) safety: Only when the trigger is pulled, the firing pin is lifted and thus within the reach of the hammer – similar to the later FEG pistols Walam 48, R61 or PA63.

Walther PP Prototype 2

Characteristics:

- Slide legend left hand side (in italics, in three lines): "Waffenfabrik Walther Zella-Mehlis (Thür.) / Walther's Patent / Polizei-Modell" – without the typical Walther banner
- No serial or project number
- Proof marks: "Crown over N"
- Sights: front and rear sight with a sighting groove in between

The following characteristics are still similar to the ones of Walther Mod. 8/9 pistols:

- Long safety lever on left hand side of frame (in front of the grip plate) meshing with the sear – cf. patent number 325 374 "Handfeuerwaffe" (handgun), filed on 24th June 1919, granted on 13th September 1920
- Magazine with a follower as a punched and bent sheet metal part, gap between end of follower and magazine body (but magazine capacity: 10 rounds in caliber 7.65 mm Browning - .32 ACP)
- Extractor with flat outside, sits half way up in ejection port
- Ejector fixed, thus without a slide hold-open function
- The following characteristics match the ones of the later Walther PP:
- Exposed hammer with hammer shaft and hammer flap
- Sear activates the hammer and is extended at the bottom to mesh with the safety lever
- Magazine release as a push button in the upper frame
- Two-piece grips made of checkered wood, with grip screw, without Walther banner

Walther PP Prototype 3

Characteristics: as prototype 2, but

- Project number "2" on right hand side of slide

Walther PP Prototype 4

Characteristics: as prototypes 2 and 3, but

- For the first time with a serial number: 750 000-1
- Slide legend left hand side (in italics, in

two lines): "Waffenfabrik Walther Zella-Mehlis (Thür.) / Walther's Patent Cal. 7,65 m/m" with Walther banner - on right hand side (!) of inscription

- No model designation
- Semi-shrouded hammer, massive hammer spur
- Extractor sits higher up in ejection port, outside contoured – cf. later PP/PPK

Walther PP Prototype 5

Characteristics: as prototype 4, but

- Serial number: 750 000-3

Walther PP Prototype 6

Characteristics: as prototype 5, but

- Serial number (in italics, electro-penciled on right hand side of slide): 421 317 (serial number fits into the Mod. 8/9 range)
- Project number (in italics): "E-4" (on left hand side of slide and on frame below the serial number); "E" might stand for "Entwicklung" (development) or "Experiment"
- Number "3" on right hand side of slide
- Semi-shrouded hammer
- Ejector no longer fixed (can be removed to the side), but no slide hold-open function yet
- Pivoting safety lever (in left hand side of slide within the serrated area); safety acts on the firing pin, as the safety shaft covers the end of the firing pin when on "safe
- Two-piece firing pin in a housing – cf. later 1st variation Walther PP
- Magazine extension made of metal, inserted into the magazine body

Picture 12: *Walther PP ZM prototype 6, SN 421 317 – E4, caliber 7.65 mm Browning (.32 ACP)*

Walther PP Prototype 7

Characteristics: as prototype 6, but

- Serial number: 750 007 (concealed on left hand side of frame, below the ejector, covered by the slide)
- Project number: "1" on the grip, but concealed
- Hammer block on top of the frame with a "C"-shaped spring – cf. later 1st variation Walther PP
- Ejector no longer fixed, but pivoting 5° upwards; thus this part for the first time serves as a slide hold-open device, when the magazine is emptied
- Hammer totally exposed; spur no longer massive, but ring-shaped
- For the first time loaded chamber indicator (so-called "signal pin") at the rear of the slide; protrudes from the slide, if there is a round in the chamber

Walther PP Prototype 8

Characteristics: as prototype 7, but

- Without a serial number
- Project number: "2" on front of grip
- Sear is no longer extended at the bottom in order to mesh with the safety lever – cf. later 1st variation Walther PP
- Rear end of trigger bar is mounted by a separate spring, which sits in a cutout in the frame – cf. later 1st variation Walther PP
- Plunger of main spring sits in the heel of the frame – cf. later Walther PP
- The front of the slide has a dovetail, conducted in a slot in the frame. This design feature was not used in the series production of the Walther PPs and PPKs, but much later in the design of the Walther PP90 – see there.

Walther PPK Prototype

After 1930 the Walther "Polizei-Pistole" (PP) achieved a considerable sales success, due to its reliable double-action system. So there was soon a demand for a shorter version of this pistol. PPs were usually carried by uniformed staff in a belt holster. Other users, who carried their service pistol concealed (e.g. plain clothes police officers), preferred a shorter pistol.

So presumably from 1930/31 onwards Walther dealt with the development of such a model. At such an early stage there was of course no new model designation given to this model yet. So the early pre-series PPKs still carried the model designation "Mod. PP" on the slide – up to about serial number 757 950.

With the beginning of the series production Walther had to deal with the naming of this new model. Often one can read the erroneous assumption that "PPK" stands for "Polizei-Pistole kurz" (short police pistol). It is correct, however, that this abbreviation stands for "Polizei-Pistole Kriminal", as the plain clothes police (so-called "Kriminalpolizei") were the target group [no joke intended!] of this shortened version of the Walther PP.

It is not clear, why this model was not named "KPP" (for "Kriminalpolizei-Pistole). One possible explanation might be that Walther – similar to the inventory designations of the modern German military – put the greatest common denominator to the front: "Pistole" – "Polizei" – "Kriminal" – hence "PPK". This is only an assumption, however, which needs evidence.

In 2011 a special Walther pistol surprisingly surfaced in the collection of a Northern German gun collector. It shows the serial number 751 405 and the model designation "Mod. PP". Despite this designation that specimen shows characteristics, which clearly belong to a Walther PPK: length of slide and barrel, one-piece wrap-around grip – though the latter are

Picture 13: *Walther PP ZM prototype 7, SN 750 007-1, caliber 7.65 mm Browning (.32 ACP)*

made of wood. The serial number, which is stamped "upside-down" (i.e. legible from the muzzle side) on the frame is within the serial number range of the 1930 PP production. So obviously we here have a PPK prototype.

Characteristics similar to early Walther PPs (1929 - 1931 production):

- Slide legend and model designation on left hand side of slide
- Serial number "upside-down" on right hand side of frame (serial numbers of later PPs/PPKs can be read from the grip panel side)
- Barrel without a collar at the rear end to secure it in the frame (barrel holder); so the barrel was pressed into the frame from the front and then secured by a pin only
- Trigger bar with a separate spring (later PPs/PPKs have a combined spring for the trigger and the trigger bar)
- Sear: both sides symmetric, semi-circular shaped seen from the side (later PPs/PPKs have an asymmetric sear)
- Hammer block with a flat top (later PPs/PPKs have a hammer block with a small dimpled dish on top for an easier dismantling of the hammer block)

Characteristics similar to Walther PPK (early series PPK production, which began around SN 760 000):

- The bottom of the slide (below the muzzle) is not stepped, but curved. The muzzle part of the PP slide was obviously cut off, shortened and welded back on. The weld seams were grinded down carefully and smoothly. The fit is very precise. Such curved fronts of the slide can still be found on early PPKs in the serial number range around 754 950 – and on post-1945 PPKs.
- The magazine release sits at the heel of the grip (similar to, but not identical with the ones used on later PPs/PPKs in caliber 9 mm kurz - .380 ACP or on PPs/PPKs for the "Polizeidirektion München"). Moreover this magazine catch is not made

Picture 14: *Walther PPK ZM prototype, SN 751 405, caliber 7,65 mm Browning (.32 ACP)*

of punched sheet metal, but milled out of steel. Obviously there was a magazine release push button (as on later PPs/PPKs) in the upper frame before, but this one was removed and the holes were professionally welded and grinded down.

- The milled rear sight is slightly smaller and more compact than the ones of PPKs from the series production. For this purpose the inserted, rear PP sight was removed from the dove tail in the slide. Then the dove tail was closed by overlay-welding and a new rear sight was milled out.
- The front sight already shows the dimple for the fluorescent night sight.
- The back of the frame is partially milled away to accommodate the one-piece wrap-around grip. The contours of this milling do not correspond with the ones of later PPKs – especially in the area of the grip screw and at the upper part of the frame.
- The bottom part of the grip was shortened and a new magazine well was welded on.
- In the 751 400 serial number range Walther PPs have a two-piece, covered firing pin, a corresponding signal pin and a so-called "fat" (i.e. wide) safety lever with seven grooves. This is not true for this pistol 751 405. It can often be observed, however, that the fragile two-piece firing pin was replaced later by the more solid one-piece one. Then the "fat" safety lever usually was replaced by the narrow type (with five grooves), too, as the latter is less fragile as well. The safety shaft belonging to the "fat" lever is more vulnerable, as it has some additional millings.
- The grip is of the one-piece type, handmade of wood with checkering, but without Walther banner.

Contemplating all these characteristics this PPK prototype (SN 751 405) can be dated 1930 or slightly later.

Walther PP Zella-Mehlis

Production:	1929 - 1945
Calibers:	7.65 mm Browning = .32 ACP (8 rounds) 9 mm kurz (9x17) = .380 ACP (7 rounds) . 22 l.r. (9 rounds) very rare: 6.35 mm Browning = .25 ACP (8 rounds)
Total Length:	6.7 in.
Barrel Length:	3.6 in.
Weight:	22.9 oz. (lightweight) or 23.7 (steel)
Rifling:	6 grooves, right-twisted
Operation:	unlocked blowback action with firing pin and exposed hammer, double action/ single action trigger mechanism

Preliminary remarks:

The Walther PP was the first pistol with a "double action/ single action" trigger mechanism (see chapter "PP/PPK prototypes), which could safely be carried with a chambered round and de-cocked.

When reading the following comments one has to bear in mind that the characteristics of the different variations overlap in some border areas, as the transitions were usually flowing. When serial numbers are given for changes, they refer to the earliest date of a change. This does not necessarily mean that a specific characteristic may not appear later any more, as Walther normally used up parts from old stock first, before they produced new ones.

The chronological sequence of technical changes during the production period was investigated with the help of the author's database. Some changes can also be dated using dated design drawings. Several copies of those are in the author's possession. Technical changes were, however, not always implemented in the production concurrently with the date of the drawing.

Characteristics:

- Slide legend on left hand side (in italics): "Waffenfabrik Walther Zella-Mehlis (Thür.) / Mod. PP" with Walther banner on left hand side of inscription
- Slide legend on right hand side: none
- Proof marks: "Crown over N", from January 1940 onwards: "Eagle over N"
- Serial number: vertically on right hand side of frame behind the grip plate; first "upside-down" (i.e. legible from the muzzle side), then legible from the grip plate side; later variations with serial number additionally on right hand side of slide
- Frame: early steel PPs with a short tang at the rear of the frame and a milled step at the base of the trigger guard, later steel PPs with a longer frame tang and no more step at the base of the trigger guard; Lightweight PP frames made of "Duraluminium" (short: "Dural"; an alloy, which was produced by Dürener Metallwerke since 1909) are rare. Early Dural frames are anodized in a dark-blue tone; later Dural frames were painted with a black lacquer.
- Fixed barrel, which serves as a guide for the recoil spring
- Barrels of early PPs pressed into the frame (barrel holder) from the front; later ones pressed in from the rear, secured by a pin
- Pivoting, spring-loaded trigger guard, which serves as a dismantling device
- Trigger freestanding, protected by trigger guard
- Sights: guilloché slide rib with milled front sight and adjustable rear sight (inserted in a dovetail)
- 22 slide serrations on both sides of the slide
- Ejection port on right hand side of slide:

early PPs with blued port, later PPs with bright metal surface in port, very late PPs with blued port again

- Exposed hammer with a round spur: early PPs without a notch on top of the spur (for the signal pin); later PPs with a notch, which makes sure that hammer and signal pin do not collide
- Hammer shaft: early PPs have a shaft with a semi-circular ridge on the inside of the left hand side flat part, later PPs have a simplified shaft without that ridge
- Hammer flap (to operate the double action mode): first wide, later narrow type
- Hammer block (blocks the movement of the hammer unless the trigger is pulled): early PPs with a flat-surfaced hammer block and a C-shaped spring, later PPs with a small dimpled dish on top (for an easier dismantling of the hammer block) and a coil spring – see illustration 1;
- Remark: In the SN-range around 773 300 – 776 800 Walther experimented with a slotted dish for the hammer block ("screw"-type).
- Hammer strut: early PPs made of milled, round steel, later PPs made of punched steel
- Main spring in the backstrap of the frame, main spring plunger (in the heel of the grip) first made of metal, later of plastic (often replaced by the lanyard ring holder – see below)
- Internal trigger bar: early PPs with blued finish, later PPs with bright metal surface finish, very late PPs with blued finish again;
- Sear: early PPs with a semi-circular, symmetrical sear (seen from the side), later PPs with an asymmetrical sear – cf. illustration 2
- Magazine release: To release the magazine from the magazine well there are two types of catches – the so-called "upper magazine release" (a spring-loaded push button on the upper left hand side of the frame) and the "bottom magazine release" (a spring-loaded latch in the heel of the grip). In the English-speaking Walther literature and in Walther forums these are often abbreviated with "UMR" and "BMR". Upper magazine release: PPs in caliber .22 l.r., .25, .32 and PPs in caliber .380 from about SN 782 304 to SN 931 250. Bottom magazine release: PPs in caliber .380 from SN 931 250 and PPs in caliber .32 for "Polizeidirektion München – PDM"
- Pivoting safety lever (in left hand side of slide within the serrated area), safety acts on the firing pin; remarks: With the early type two-piece firing pin the safety shaft covers the end of the firing pin when in the position "safe". With the later one-piece firing pin the safety shaft meshes with the one or two lugs at the rear of the firing pin when in the position "safe"
- Width of safety lever: early PPs with so-called "fat" lever (7 grooves), later PPs with "slim" lever (5 grooves)
- Swiveling range of safety lever: first 90°, later 60° - with a "flowing" transition, i.e. the last PP with 90° safety was SN 967 700 and the first PP with 60° safety was SN 948 950; so in 1936/37 between 948 950 and 967 700 both types were used
- Finish of safety lever: early PPs with matte-blued finish, later PPs with bright-blue finish, very late PPs with matte-blued finish again
- Patent number 390 298 (see above) claiming protection for "Handfeuerwaffe mit Hahnschloss" (hand gun with hammer breech action) includes a grip safety, too. But this was not realized in Mod. PP.
- Firing pin: early PPs with a two-piece firing pin in a U-shaped housing (see illustration 1); later PPs with a one-piece firing pin in a U-shaped housing; then PPs with a one-piece firing pin without a housing, underside of slide milled through to get access to the firing pin and the signal pin; with PPs in caliber .22 l.r. this underside is not milled through, however,

as there cannot be a signal pin because of the rimfire cartridge; remarks: In the SN-range around 761 000 – 762 600 Walther experimented with a one-piece firing-pin with one lug on either side at the rear. Standard one-piece firing pins have one lateral lug at the rear end only.

- Signal pin (i.e. loaded chamber indicator), which protrudes from the rear of the slide indicating, that there is a round in the chamber (optional feature at an extra charge)
- Ejection port: early PPs with blued port, later PPs with bright metal surface in port, very late PPs with blued port again
- Exposed extractor, in ejection port, outside contoured
- Ejector pivoting 5° upwards, thus this part works as a slide hold-open device, too, when the magazine is emptied; finish: early PPs blued, later PPs bright metal finish, very late PPs blued again; ejectors for PPs in caliber .380 have an additional "spur" at the bottom
- Magazine follower: with early PPs as a punched and bent sheet metal part, gap between end of follower and magazine body; with later PPs as a punched and deep-drawn sheet metal part, end of follower is folded into the magazine body; with some PPs in caliber .380 as a part which is milled out of solid steel
- Magazine floor plate: inserted into magazine body – either with a magazine extension or without one; magazine extensions made of brown, black, green or red plastic – see above for details
- Metal finish: early PPs with matte-blued finish, later PPs with high-gloss polished finish, very late PPs with dull "military" finish (from 1941/42 by order of the "Reichsminister für Bewaffnung und Munition"); matte chrome-plated, gold-, silver- or nickel-plated finish or engravings (with blued, gold-, silver- or nickel-plated finish) optional at an extra charge; remark: High-gloss chrome-plated PPKs are presumably not authentic, but were plated later.
- Grip plates: with early PPs made of black, checkered, hard rubber; with later PPs made of "Trolitan" (see above); ivory or

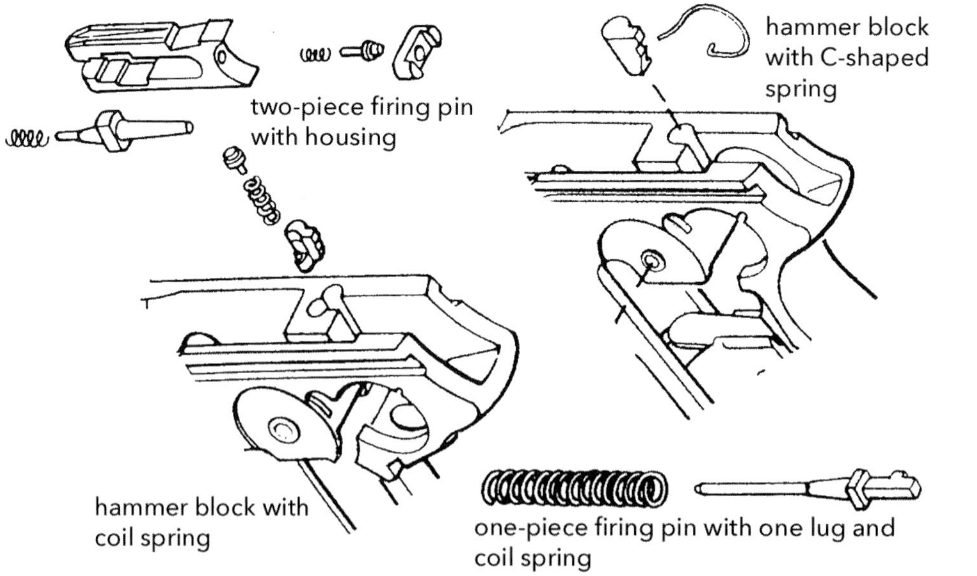

Illustration 1: *firing pins and hammer blocks PP/PPK (according to E.J. Hoffschmidt see bibliography)*

carved wood grips optional at an extra charge
- Lanyard ring: optional item (when installed it replaces the main spring plunger) with a swiveling ring – e.g. on most Swedish police PPs; fixed lanyard rings, which sit sideways at the heel of the grip in a cutout within the grip plates, are usually later additions – e.g. by French armorers.

Walther PP Zella-Mehlis Development and Technical Changes

Mod. PP, 1st Variation (1929 – end of 1930)

SN 750 050 to 757 000

Characteristics:
- Slide legend left hand side (in italics): "Waffenfabrik Walther Zella-Mehlis (Thür.) / Mod. PP" with Walther banner on left hand side of inscription
- Slide legend right hand side: none
- Proof marks: "Crown over N"
- Serial number: vertically on right hand side of frame behind grip plate; "upside-down" (i.e. legible from the muzzle side)
- Frame: with a short tang at the rear of the frame and a milled step at the base of the trigger guard
- Fixed barrel, which serves as a guide for the recoil spring
- Barrel pressed into the frame (barrel holder) from the front, secured by a pin
- Pivoting, spring-loaded trigger guard, which serves as a dismantling device
- Trigger freestanding, protected by trigger guard
- Sights: guilloché slide rib with milled front sight and adjustable rear sight (inserted in a dovetail)
- 22 slide serrations on both sides of the slide
- Ejection port on right side of slide with blued port
- Exposed hammer with a round spur, without a notch on top of the spur (for the signal pin), so-called "heavy hammer" (i.e. only a 4 mm lightening hole in the middle)
- Hammer shaft: with a semi-circular ridge on the inside of the left hand side flat part
- Hammer flap (to operate the double action mode): wide type
- Hammer block (blocks the forward movement of the hammer unless the trigger is pulled) with a flat-surfaced hammer block and a C-shaped spring – see illustration 1
- Hammer strut: made of milled, round steel
- Main spring in the backstrap of the frame, main spring plunger (in the heel of the grip) made of metal (often replaced by the lanyard ring holder)
- Internal trigger bar with blued finish; rear end of trigger bar is mounted by a separate spring, which sits in a cutout in the frame
- Sear: with semi-circular, symmetrical sear (seen from the side), fixed to frame with two separate little pins
- Magazine release: push button on upper left hand side of frame
- Pivoting safety lever (in left hand side

Picture 15: *Walther PP ZM, 1st variation (1930), SN 753 373, caliber 7.65 mm Browning (.32 ACP), 90°-safety with "fat" safety lever*

of slide within the serrated area), safety acts on the firing pin, as the safety shaft covers the end of the firing pin when in the position "safe"

- Width of safety lever: so-called "fat" lever (7 grooves)
- Swiveling range of safety lever: 90°
- Finish of safety lever: matte-blued finish
- Firing pin: up to about SN 757 500 mostly two-piece firing pin in a U-shaped housing (see illustration 1), some pistols subsequently fitted with the one-piece firing pin; from about SN 751 740 some pistols already with the one-piece firing pin in a U-shaped housing
- Signal pin (i.e. loaded chamber indicator): spring loaded, round pin with a rectangular collar in the middle to sit on top of the firing pin housing
- Ejection port: blued
- Exposed extractor: in ejection port, outside contoured
- Ejector pivoting 5° upwards, thus this part works as a slide hold-open device, too, when the magazine is emptied; finish: blued
- Magazine follower: as a punched and bent sheet metal part, gap between end of follower and magazine body
- Magazine floor plate: inserted into magazine body – either with a magazine extension or without one
- Metal finish: matte-blued finish; matte chrome-plated, gold-, silver- or nickel-plated finish or engravings (with blued, gold-, silver- or nickel-plated finish) optional at an extra charge
- Grip plates: black, checkered, hard rubber; with a big X-shaped reinforcement inside; ivory or carved wood grips optional at an extra charge
- Lanyard ring: optional item (when installed it replaces the main spring plunger) with a swiveling ring – e.g. on most Swedish police PPs; fixed lanyard rings, which sit sideways at the heel of the grip in a cutout within the grip plates, are usually later

additions – e.g. by French armorers.

Remarks:

As the combination of the two-piece firing pin (in a housing) and the pivoting safety with the "fat" lever is very fragile, due to additional millings, many Walther PPs were converted to the later type – i.e. covered one-piece firing pin with a safety with a "slim" lever. This was done e.g. with most "Reichsbank" PPs.

Mod. PP, 2nd Variation (end of 1930 – mid-1932)

SN 757 000 to 769 000

Characteristics: as before, but

- Serial number: vertical on right hand side of frame behind grip plate; from about SN 767 500 mostly no longer "upside-down" (i.e. legible from the muzzle side), but legible from the grip plate side
- Exposed hammer with a round spur; mostly with a so-called "big hammer ring" (i.e. a 6 mm lightening hole in the middle with slightly chamfered edges)
- Hammer block (blocks the forward movement of the hammer unless the trigger is pulled): with a flat surface and a C-shaped spring; according to a design drawing dated 5th December 1931 from then onwards with a small dimpled dish on top (for an easier dismantling of the hammer block) and a coil spring – see illustration 1
- Width of safety lever: "slim" lever (5 grooves)
- Swiveling range of safety lever: 90°
- Finish of safety lever: matte-blued
- Firing pin: up to about SN 757 500: some pistols still with the two-piece firing pin (in a U-shaped housing); however: most pistols with a one-piece firing pin (in a U-shaped housing); from about SN 765 000: most pistols with a one-piece firing pin without a housing (underside of slide milled through to get access to the firing pin and the signal pin; with PPs in caliber .22 l.r. this underside is not milled through, however, as there cannot be a signal pin because of the rimfire cartridge); [format?] remarks: In the SN-range around 761 000 – 762 600 Walther experimented with a one-piece firing-pin with one lug on either side at the rear. Standard one-piece firing pins have one lateral lug at the rear end only.
- Signal pin (i.e. loaded chamber indicator): spring loaded, round pin with a rectangular collar in the middle to sit on top of the firing pin housing
- Magazine follower: punched sheet metal part, end of follower is folded into the magazine body; with some PPs in caliber .380 as a part milled out of solid steel
- Metal finish: matte-blued finish or high-gloss polished finish; matte chrome-plated, gold-, silver- or nickel-plated finish or engravings (with blued, gold-, silver- or nickel-plated finish) optional at an extra charge
- Grip plates: black, checkered, hard rubber with a big or a thin X-shaped reinforcement inside

Mod. PP, 3rd Variation (mid-1932 – 1935)

SN 769 000 to 948 000

Characteristics: as before, but

- Serial number: vertically on right hand side of frame behind grip plate; no longer "upside-down" (i.e. legible from the muzzle side), but legible from the grip plate side
- Barrel pressed in from the rear, secured by a collar and a pin (from 12th July 1932 - according to design drawing)
- Exposed hammer with a round spur and a notch on top, which makes sure that hammer and signal pin do not collide; hammer spur either "big hammer ring" type or standard type (i.e. with a 4 mm

chamfered lightening hole in the middle)

- Hammer block with a small dimpled dish on top for an easier dismantling of the hammer block and a coil spring – see illustration 1; remark: In the SN-range around 773 300 – 776 800 Walther experimented with a slotted dish for the hammer block ("screw"-type).
- Trigger bar: up to about SN 776 000 with a separate spring at the rear end, from about SN 776 000 a combined trigger/trigger bar spring takes over this function
- Sear: asymmetrical; from about SN 776 600 fitted to frame by pins, which are firmly mounted on the sear
- Magazine release: push button on upper left hand side of frame (PPs in caliber .22 l.r., .25, .32 and PPs in caliber .380 before SN 931 244) or bottom magazine release with a latch in the heel of the grip (later PPs in caliber .380)
- Width of safety lever: "slim" lever (5 grooves)
- Swiveling range of safety lever: 90°
- Signal pin (i.e. loaded chamber indicator): spring loaded pin with a rectangular middle part, dimple at the underside for easy removal from slide
- Metal finish: mostly high-gloss polished finish; matte chrome-plated, gold-, silver- or nickel-plated finish or engravings (with blued, gold-, silver- or nickel-plated finish) optional at an extra charge
- Grip plates: either black, checkered, hard rubber with a thin X-shaped reinforcement inside or with cast-in metal reinforcements (in the US-literature often named "weighted grips". This designation is simply wrong, as these metal plates are nothing but reinforcements); outside checkered, with Walther banner

Mod. PP, 4th Variation (1935 – end of 1941)

SN 948 000 to 223 000P

Characteristics: as before, but

- Serial number: suffix "P" added from SN 100 000P – see above
- Proof marks: from about January 1940 "Eagle over N", instead of "Crown over N" (but flowing transition!)
- Swiveling range of safety lever: 60° (flowing transition – see above)
- Safety lever mostly fire blued
- Hammer strut: made of punched steel from about SN 151 000P (November 1939)
- Hammer flap (to operate the double action mode): narrow type from about SN 151 000P (November 1939)
- Grip plates: either made of hard rubber with cast-in metal reinforcements (in the US-literature often named "weighted grips". This designation is simply wrong, as these metal plates are nothing but reinforcements) or made of black or reddish brown "Trolitan" with a cavity inside ("Trolitan" is an injection-molded or pressed plastic based on phenolic resin – similar to "Bakelite". According to Walther design drawings those grip plates should have been used from 1st September 1938 already); both types outside checkered, with Walther banner

Mod. PP, 5th Variation (end of 1941 – January 1942)

SN 223 000P to 232 200P

Characteristics: as before, but

- Frame: with a long tang at the rear of the frame and still a milled step at the base of the trigger guard (not true for "Dural" frames, which kept the short frame tang)

Picture 16: *Walther PP ZM, 4th variation (1938), SN 993 147, caliber 7.65 mm Browning (.32 ACP), 60°-safety with "slim" safety lever, NSKK-emblem*

Mod. PP, 6th Variation (January 1942 – April 1945)

SN 232 200P to 397 000P

Characteristics: as before, but

- Slide legend: from about SN 368 800P (end of 1944) left hand side inscription omitted due to secrecy rules, "ac"-code stamp (for: Waffenfabrik Carl Walther/ Zella-Mehlis) on right hand side of slide instead
- Frame: no more milled step at the base of the trigger guard (from 26th January 1942 – according to design drawing, but flowing transition); long tang at the rear of the frame (not true for "Dural" frames, which kept the short frame tang)
- Serial number from 1942 additionally on right hand side of slide from SN 247 300P; from about SN 383 000P serial number on the slide often no longer matching the one on the frame;
- Remark: PPs with the serial number on right hand side of slide before SN 247 300P were either for special contracts (e.g. "Reichssicherheitshauptamt", "RSHA" – see there) or the serial number was stamped or engraved later (e.g. by East German armed forces – see there)
- Signal pin (i.e. loaded chamber indicator): spring loaded pin with "waisted" middle part, dimple at the underside for easy removal from slide – from June 1942
- Proof marks: from about SN 383 000P "Eagle over N" proof mark mostly no longer stamped
- Hammer shaft: from about SN 277 500P (1942) simplified shaft without the semi-circular ridge on the inside of the left hand side flat part
- Magazines: from about SN 343 000P stamp "w" on right hand side of magazine body ("w" for "weich" – i.e. not hardened); from 11th June 1942 - according to the design drawing, but obviously not implemented before 1944
- Dull "military" finish, as the polishing

Picture 17: *Walther PP ZM, 6th variation (1945), SN 386 486P, caliber 7.65 mm Browning (.32 ACP), "ac"-code, grips made of pressed wood*

before the blueing process was omitted (from 1941/42 by order of the "Reichsminister für Bewaffnung und Munition")

- Grip plates: from about SN 350 480P (1944) grip plates often made of pressed beech wood with Walther banner (but flowing transition!)

Total production of Zella-Mehlis Walther Mod. PP pistols: about 540,000 pistols

Walther PPK Zella-Mehlis

Production:	1931 - 1944/45
Calibers:	7.65 mm Browning = .32 ACP (7 rounds) 9 mm kurz (9x17) = .380 ACP (7 rounds) . 22 l.r. (9 rounds) very rare: 6.35 mm Browning = .25 ACP (8 rounds)
Total Length:	6.1 in.
Barrel Length:	3.3 in.
Weight:	16.9 oz. (lightweight) or 19.2 (steel)
Rifling:	6 grooves, right-twisted
Operation:	unlocked blowback action with firing pin and exposed hammer, double action/ single action trigger mechanism

Preliminary remarks:

The Walther PPK is the shortened version of the Walther PP. The letter "K" does not stand for "kurz" (short) though, but for "Kriminal" (see above).

When reading the following comments one has to bear in mind that the characteristics of the different variations overlap in some border areas. When serial numbers are given for changes, they refer to the beginning of a change. This does not necessarily mean that a specific characteristic may not appear later any more, as Walther normally used up parts from old stock first, before they produced new ones.

The chronological sequence of technical changes during the production period was determined using the author's database. Some changes can also be dated with the aid of dated design drawings. Several copies of those are in the author's possession. Technical changes were, however, not always implemented in the production concurrently with the date of the drawing.

Characteristics:

- Slide legend on left hand side (in italics): "Waffenfabrik Walther Zella-Mehlis (Thür.) / Mod. PPK" with Walther banner on left hand side of inscription; remarks: from about SN 754 950 – 757 950 Walther PPK were still designated "Mod. PP"; from SN 191 156K – 191 192K: Walther PPKs erroneously designated "Mod. PP" on slide again
- Slide legend on right hand side: none
- Proof marks: "Crown over N", from January 1940: "Eagle over N"
- Serial number: vertically on right hand side of frame behind grip plate; first "upside-down" (i.e. legible from the muzzle side), then legible from the grip plate side; later variations with serial number additionally on right hand side of slide
- Frame: early steel PPKs with a short tang at the rear of the frame and a milled step at the base of the trigger guard, later steel PPKs with a longer frame tang and no more step at the base of the trigger guard; Lightweight PPK frames made of "Duraluminium" (short: "Dural"; an alloy, which was produced by Dürener Metallwerke since 1909) are rare. Early Dural frames are anodized in a dark-blue tone, later Dural frames were painted with a black lacquer.
- Slides: Zella-Mehlis made PPKs have a stepped contour at the bottom of the slide (below the muzzle); except for PPK prototypes and very early PPKs – see there).
- Fixed barrel, which serves as a guide for the recoil spring
- Barrels of early PPKs pressed into the frame (barrel holder) from the front, later ones pressed in from the rear, secured by a pin
- Pivoting, spring-loaded trigger guard, which serves as a dismantling device
- Trigger freestanding, protected by trigger guard
- Sights: guilloché slide rib with milled front sight, rear sight first milled as an integral part of the slide rib, later adjustable PP-type rear sight (inserted in a dovetail)
- 22 slide serrations on both sides of the slide
- Ejection port on right hand side of slide: early PPKs with blued port, later PPKs with bright metal surface in port, very late PPKs with blued port again
- Exposed hammer with a round spur: early PPKs without a notch on top of the spur (for the signal pin); later PPKs with a notch, which makes sure that hammer and signal pin do not collide
- Hammer shaft: early PPKs have a shaft with a semi-circular ridge on the inside of the left hand side flat part, later PPKs have a simplified shaft without that ridge
- Hammer flap (to operate the double action mode): first wide, later narrow type
- Hammer block (blocks the forward

movement of the hammer unless the trigger is pulled): early PPKs with a flat-surfaced hammer block and a C-shaped spring, later PPKs with a small dimpled dish on top (for an easier dismantling of the hammer block) and a coil spring – see illustration 1; remark: In the SN-range around 774 740 – 776 450 Walther experimented with a slotted dish for the hammer block ("screw"-type).

- Hammer strut: early PPKs made of milled, round steel; later PPKs made of punched steel
- Main spring in the backstrap of the frame, main spring counter bearing at the heel of the grip serves as a fixed lanyard loop
- Internal trigger bar: early PPKs with blued finish, later PPKs with bright metal surface finish, very late PPKs with blued finish again
- Sear: early PPKs with semi-circular, symmetrical sear (seen from the side), later PPKs with an asymmetrical sear
- Magazine release: To release the magazine from the magazine well there are two types of catches – the so-called "upper magazine release" (a spring-loaded push button on the upper left hand side of the frame) and the "bottom magazine release" (a spring-loaded latch in the heel of the grip). In the English-speaking Walther literature and in Walther forums these are often abbreviated with "UMR" and "BMR". Upper magazine release: PPKs in caliber .22 l.r., .25, .32 and PPKs in caliber .380 from SN 795 900 to about SN 1 001 550. Bottom magazine release: PPKs in caliber .380 from SN 1 001 550 and PPKs in caliber .32 for "Polizeidirektion München – PDM"
- Pivoting safety lever (in left hand side of slide within the serrated area), safety acts on the firing pin; remarks: With the early type two-piece firing pin the safety shaft covers the end of the firing pin when in the position "safe". With the later one-piece firing pin the safety shaft meshes with the one or two lugs at the rear end of the firing pin when in the position "safe".
- Width of safety lever: "slim" lever (5 grooves), no examples with "fat" levers (7 grooves) were reported so far
- Swiveling range of safety lever: first 90°, later 60° - with a "flowing" transition, i.e. the last PPK with 90° safety was SN 999 795 and first PPK with 60° safety was SN 853 020; so in 1936/37 between 853 020 and 999 795 both types were used
- Finish of safety lever: early PPKs with matte-blued finish, later PPKs with bright-blue finish, very late PPKs with matte-blued finish again
- Patent number 390 298 (see above) claiming protection for "Handfeuerwaffe mit Hahnschloss" (hand gun with hammer breech action) includes a grip safety, too. But this was not realized in Mod. PPK.
- Firing pin: early PPKs with a two-piece firing pin in a U-shaped housing (see illustration 1); later PPKs with a one-piece firing pin in a U-shaped housing; then PPKs with a one-piece firing pin without a housing, underside of slide milled through to get access to the firing pin and the signal pin; with PPKs in caliber .22 l.r. this underside is not milled through, however, as there cannot be a signal pin because of the rimfire cartridge; remarks: There are hardly any PPKs reported with the two-piece firing pin and the "fat" safety. Presumably most of these were "upgraded" to the one-piece firing pin type. In the SN-range around 760 800 – 762 000 Walther experimented with a one-piece firing-pin with one lug on either side at the rear. Standard one-piece firing pins have one lateral lug at the rear end only.
- Signal pin (i.e. loaded chamber indicator), which protrudes from the rear of the slide, if there is a round in the chamber (optional feature at an extra charge)
- Ejection port: early PPKs with blued port, later PPKs with bright metal surface in port, very late PPKs with blued port again
- Exposed extractor, in ejection port, outside

contoured

- Ejector pivoting 5° upwards, thus this part works as a slide hold-open device, too, when the magazine is emptied; finish: early PPKs blued, later PPKs bright metal finish, very late PPKs blued again; ejectors for PPKs in caliber .380 have an additional "spur" at the bottom
- Magazine follower: with early PPKs as a punched and bent sheet metal part, gap between end of follower and magazine body; with later PPKs as a punched and deep-drawn sheet metal part, end of follower is folded into the magazine body; with some PPKs in caliber .380 as a part milled out of solid steel
- Magazine floor plate: inserted into magazine body – either with a magazine extension or without one; magazine extensions made of brown, black, green or red plastic – see above for details
- Lanyard loop: punched and bent steel item at the heel of the grip, serves as the main spring counter bearing
- Metal finish: early PPKs with matte-blued finish, later PPKs with high-gloss polished finish, very late with dull "military" finish (from 1941/42 by order of the "Reichsminister für Bewaffnung und Munition"); matte chrome-plated, gold-, silver- or nickel-plated finish or engravings (with blued, gold-, silver- or nickel-plated finish) optional at an extra charge; remark: High-gloss chrome-plated PPKs are presumably not authentic, but were plated later.
- Grips: one-piece wrap-around type grips made of mottled brown "Trolit" ("Trolit" is an injection-molded plastic based on cellulose acetate); outside checkered, with Walther banner; first grips without a cast-in metal reinforcement in the lower backstrap, then with one; grip screw first from the left to the right; later from the right to the left; color of plastic material changing from mottled brown to chocolate brown, then grey-brown and finally black

Walther PPK Zella-Mehlis Development and Technical Changes

Mod. PPK, 1st Variation (beginning of 1931 – mid-1931)

SN 754 950 to 760 000

Characteristics:

- Slide legend left hand side (in italics): "Waffenfabrik Walther Zella-Mehlis (Thür.) / Mod. PPK" with Walther banner on left hand side of inscription; remark: from about SN 754 950 – 757 950 Walther PPK were still designated "Mod. PP"
- Slide legend right hand side: none
- Proof marks: "Crown over N"
- Serial number: vertically on right hand side of frame behind grip plate; "upside-down" (i.e. legible from the muzzle side)
- Frame: with a short tang at the rear of the frame and a milled step at the base of the trigger guard
- Slides: very early Zella-Mehlis made PPKs still have a curved contour at the bottom of the slide (below the muzzle); series production PPK slides have a stepped contour there (cf. PPK prototype and post-1945 PPKs – see there).

- Fixed barrel, which serves as a guide for the recoil spring
- Barrel pressed into the frame (barrel holder) from the front, secured by a pin and a collar
- Pivoting, spring-loaded trigger guard, which serves as a dismantling device
- Trigger freestanding, protected by trigger guard
- Sights: guilloché slide rib with milled front and rear sight as an integral part of the slide rib
- 22 slide serrations on both sides of the slide
- Ejection port on right hand side of slide with blued port
- Exposed hammer with a round spur, without a notch on top of the spur (for signal pin)
- Hammer shaft: with a semi-circular ridge on the inside of the left hand side flat part
- Hammer flap (to operate the double action mode): wide type
- Hammer block (blocks the forward movement of the hammer unless the trigger is pulled) with a flat-surfaced hammer block and a C-shaped spring – see illustration 1
- Hammer strut: made of milled, round steel
- Main spring in the backstrap of the frame, main spring counterbearing at the heel of the grip serves as a fixed lanyard loop
- Internal trigger bar with blued finish; rear end of trigger bar is mounted by a separate spring, which sits in a cutout in the frame
- Sear: semi-circular, symmetrical (seen from the side), fixed to frame with two separate little pins
- Magazine release: push button on upper left hand side of frame
- Pivoting safety lever (in left hand side of slide within the serrated area), safety acts on the firing pin, as the safety shaft covers the end of the firing pin when in the position "safe"
- Width of safety lever: so-called "fat" lever (7 grooves)
- Swiveling range of safety lever: 90°
- Finish of safety lever: matte-blued finish
- Firing pin: mostly two-piece firing pin in a U-shaped housing (see illustration 1), some pistols subsequently fitted with the one-piece firing pin (then often with two lugs at the rear end); from about SN 757 910: most pistols with the one-piece firing pin (in a U-shaped housing)
- Signal pin (i.e. loaded chamber indicator): spring loaded, round pin with a rectangular collar in the middle to sit on top of the firing pin housing
- Ejection port: blued
- Exposed extractor: in ejection port, outside contoured
- Ejector pivoting 5° upwards, thus this part works as a slide hold-open device, too, when the magazine is emptied; finish: blued
- Magazine follower: as a punched and bent sheet metal part, gap between end of follower and magazine body
- Magazine floor plate: inserted into magazine body – either with a magazine extension or without one
- Metal finish: matte-blued finish; matte chrome-plated, gold-, silver- or nickel-plated finish or engravings (with blued, gold-, silver- or nickel-plated finish) optional at an extra charge
- Grips: one-piece wrap-around type grips made of mottled brown "Trolit" ("Trolit" is an injection-molded plastic based on cellulose acetate); outside checkered, with Walther banner; without a cast-in metal reinforcement in the lower backstrap; grip screw from the left to the right; color of plastic material: mottled brown

Mod. PPK, 2nd Variation (mid-1931 – mid-1932)

SN 760 000 to 774 500

Picture 18: *Walther PPK ZM, 2nd variation (1932), SN 759 444, caliber 7.65 mm Browning (.32 ACP), 90°-safety*

Characteristics: as before, but

- Serial number: vertically on right hand side of frame behind grip plate; from about SN 767 450 no longer "upside-down" (i.e. legible from the muzzle side), but legible from the grip plate side
- Exposed hammer with a round spur; mostly with the so-called "big hammer ring" (i.e. a 6 mm lightening hole in the middle with slightly chamfered edges); some of the hammers with a recess milled out at the rear ("light" hammer?)
- Hammer block (blocks the forward movement of the hammer unless the trigger is pulled): still with a flat surface and a C-shaped spring; according to a design drawing dated 5th December 1931 from then onwards with a small dimpled dish on top (for an easier dismantling of the hammer block) and a coil spring – see illustration 1;
- Width of safety lever: "slim" lever (5 grooves)
- Swiveling range of safety lever: 90°
- Finish of safety lever: matte-blued
- Firing pin: up to about SN 767 000 some pistols still with the two-piece firing pin (in a housing – see illustration 1); from about SN 757 910: most pistols with the one-piece firing pin in a U-shaped housing; from SN 767 000: most pistols with the one-piece firing pin without a housing, underside of slide is milled through to get access to the firing pin and the signal pin; with PPKs in caliber .22 l.r. this underside is not milled through, however, as there cannot be a signal pin because of the rimfire cartridge; remarks: In the SN-range around 760 800 – 763 000 Walther experimented with a one-piece firing-pin with one lug on either side at the rear. Standard one-piece firing pins have one lateral lug at the rear end only.
- Trigger bar: up to about SN 776 400 with a separate spring at the rear end, from about SN 776 400 a combined trigger/trigger bar

spring takes over this function

- Signal pin (i.e. loaded chamber indicator): spring loaded, round pin with a rectangular collar in the middle to sit on top of the firing pin housing
- Magazine follower: punched and deep-drawn sheet metal part, end of follower is folded into the magazine body; with some PPKs in caliber .380 as a part milled out of solid steel
- Metal finish: matte-blued finish or high-gloss polished finish
- Grips: one-piece wrap-around type grips made of mottled brown "Trolit" ("Trolit" is an injection-molded plastic based on cellulose acetate); outside checkered, with Walther banner; without a cast-in metal reinforcement in the lower backstrap: up to about SN 765 000, with reinforcement: from about SN 765 000; grip screw from the left to the right; color of plastic material: mottled brown

Remark:

From SN 760 000 onwards one can speak of a series production of the Walther PPK, as before there were constant changes.

Mod. PPK, 3rd Variation (mid-1932 – 1935)

SN 774 500 to 853 000

Characteristics: as before, but

- Barrel pressed in from the rear, secured by a collar and a pin (from 12th July 1932 - according to design drawing)
- Exposed hammer with a round spur and a notch on top, which makes sure that hammer and signal pin do not collide; hammer spur either "big hammer ring" type or standard type (i.e. with a 4 mm chamfered lightening hole in the middle)
- Hammer block with a small dimpled dish on top (for an easier dismantling of the hammer block) and a coil spring – see illustration 1; remark: In the SN-range around 774 740 – 776 450 Walther experimented with a slotted dish for the hammer block ("screw"-type).
- Sear: asymmetrical; from about SN 797 100 fitted to frame by pins, which are firmly mounted on sear
- Magazine release: push button on upper left hand side of frame (PPKs in caliber .22 l.r., .25, .32 and PPKs in caliber .380 before SN 931 244) or bottom magazine release with a latch in the heel of the grip (later PPKs in caliber .380)
- Width of safety lever: "slim" lever (5 grooves)
- Swiveling range of safety lever: 90°
- Firing pin: solely one-piece firing pin without a housing; underside of slide is milled through to get access to the firing pin and the signal pin; with PPKs in caliber .22 l.r. this underside is not milled through, however, as there cannot be a signal pin because of the rimfire cartridge
- Signal pin (i.e. loaded chamber indicator): spring loaded pin with a rectangular middle part, dimple at the underside for easy removal from slide
- Grips: one-piece wrap-around type grips made of mottled brown "Trolit"; outside checkered, with Walther banner; grips with a cast-in metal reinforcement in the lower backstrap; grip screw from the left to the right: up to about SN 840 000, from the right to the left: from about SN 840 000; color of plastic material mottled brown
- SN-range: 821 500 – 844 470: massive block of PPKs made for "Reichszeugmeisterei" ("RZM") – see there

Picture 19: *Walther PPK ZM, 4th variation (1940), SN 264 193K, caliber 7.65 mm Browning (.32 ACP), 60°-safety, "Ehrenwaffe für politische Leiter" with fluorescent front sight*

Mod. PPK, 4th Variation (1935 – end of 1941)

SN 853 000 to 327 500K

Characteristics: as before, but

- Serial number: suffix "K" added from SN 172 000K – see above
- Proof marks: from about January 1940 "Eagle over N", instead of "Crown over N" (but flowing transition!)
- Swiveling range of safety lever: 60° instead of 90° (flowing transition – see above)
- Safety lever mostly fire blued
- Hammer strut: made of punched steel from about SN 222 000K (November 1939)
- Hammer flap (to operate the double action mode): narrow type from about SN 249 000K (November 1939)
- Metal finish: most PPKs still with high-gloss polished finish, but some already with dull "military" finish (from 1941/42 by order of the "Reichsminister für Bewaffnung und Munition")
- Grips: one-piece wrap-around type grips made of mottled brown "Trolit" ("Trolit" is an injection-molded plastic based on cellulose acetate); outside checkered, with Walther banner; with a cast-in metal reinforcement in the lower backstrap; grip screw from the right to the left; color of plastic material: mottled brown
- SN 191 156K – 191 192K: erroneously designated "Mod. PP" on slide

Mod. PPK, 5th Variation (end of 1941 – January 1942)

SN 327 500K to 356 200K

Characteristics: as before, but

- Frame: with a long tang at the rear of the frame and still a milled step at the base of the trigger guard (not true for "Dural" frames, which kept the short tang)

Picture 20: *Walther PPK ZM, 6th variation (1944), SN 423 431K, caliber 7.65 mm Browning (.32 ACP), "Waffenamt" acceptance stamp, PP-type rear sight and black grips*

- Serial number in series production additionally on right hand side of slide – from SN 334 500K; remarks: PPKs with the serial number on right hand side of slide before SN 334 500K were either for special contracts (e.g. "Reichssicherheitshauptamt", "RSHA" – see there) or the serial number was stamped or engraved later (e.g. by East German armed forces – see there)
- Metal finish: solely dull "military" finish as the polishing before the blueing process was omitted (from 1941/42 by order of the "Reichsminister für Bewaffnung und Munition")
- Grips: one-piece wrap-around type grips made of mottled brown "Trolit" ("Trolit" is an injection-molded plastic based on cellulose acetate); outside checkered, with Walther banner; with a cast-in metal reinforcement in the lower backstrap; grip screw from the right to the left; color of plastic material: often chocolate brown, greyish brown or black

Mod. PPK, 6th Variation (January 1942 – 1944/45?)

SN 356 200K to 431 097K resp. 433 521A

Characteristics: as before, but

- Frame: no more milled step at the base of the trigger guard (from 26th January 1942 – according to design drawing, but flowing transition); long tang at the rear of the frame (not true for "Dural" frames, which kept the short frame tang)
- Hammer shaft: from about SN 371 600K (June 1942) simplified shaft without the semi-circular ridge on the inside of the left hand side flat part
- Signal pin (i.e. loaded chamber indicator): spring loaded pin with a "waisted" middle part, dimple at the underside for easy removal from slide – from June 1942
- Sights: guilloché slide rib with milled front sight, adjustable PP-type rear sight

(inserted in a dovetail) – from about SN 392 800K (1942)

- Grips: one-piece wrap-around type grips made of mottled brown "Trolit" ("Trolit" is an injection-molded plastic based on cellulose acetate); outside checkered, with Walther banner; with a cast-in metal reinforcement in the lower backstrap; grip screw from the right to the left; color of plastic material: mostly black, sometimes greyish or chocolate brown
- From about 417 150K to 417 385K: mostly Dural frame and high-polish chrome-plated slide
- From about 430 600K to 430 960K: mostly in caliber .22 l.r.
- From about 430 955K to 431 097K: mostly Dural frame, nickel- or chrome-plated slide; often without proof marks
- SN 433 521A: presumably the last PPK ever produced, with the following characteristics: Dural frame in-the-white, slide refurbished and then blued, "PP-type" rear sight, black, one-piece grips.

The reason for the suffix "A" instead of "K" is unknown.

Remarks:

An outstanding survey of the development and the technical changes of the different variations of the Walther PP/PPK pistols with exploded view drawings can be found in E.J. Hoffschmidt, "Know Your Walther PP & PPK Pistols, Owner's Manual", Blacksmith Corp, Chino Valley, Az. 1975.

Total production of Zella-Mehlis Walther Mod. PPK pistols: about 290,900 pistols

Walther PP/PPK Zella-Mehlis Peculiarities, Contracts and Special Models

Design features of Walther PP/PPK pistols in the following calibers:

Caliber 9 mm kurz (.380 ACP): The first PPs in this caliber often have slides where the caliber designation "7,65 m/m" is overstamped with "9 m/m" or ".380".

Early PPs/PPKs from SN 887 000 to 931 250 have two little horns at the top of the chamber, which correspond with two little cutouts in the action face of the slide. This is supposed to improve the cartridge feeding. Other than that the dimensions of the cartridge feed are identical with the ones in other calibers.

The magazine follower with early PPs/PPKs is made of punched sheet metal. Later PPS/PPKs have followers, which were milled out of solid steel.

Early PPs (from about SN 782 304 to 931 250) and PPKs (from about SN 795 900 to 1 001 550) in this caliber have a magazine release as a push button on the upper left hand side of the frame. Later PPs/PPKs have a latch in the heel of the grip (so-called bottom magazine release).

From PP SN 885 830 to 931 250 Walther obviously produced both variations simultaneously.

According to a design drawing dated 14th

March 1938 the ejector of PPs/PPKs in caliber 9 mm kurz have an additional "spur" at the bottom – seen from the side.

The recoil spring is more powerful than the ones in other calibers.

Due to the heavier impact of the 9 mm ammunition the frames of the PPs/PPKs are reinforced: With early PPs/PPKs the symmetrical sear can be inserted into the frame through a slot in the frame, which is milled through – the so-called "open frame".

With 9 mm PPs/PPKs this slot is no longer milled through and thus the frame remains "closed" here (cf. illustration 2). The sear is altered, so that it is asymmetrical (i.e. slimmer on the left hand side) and can be inserted from the right hand side into the "closed" frame.

Thus it was possible as well to produce PPs/PPKs in this powerful caliber with Dural frames.

These "closed" frames were later taken over for Walther PPs/PPKs in other calibers, too.

Walther PPs in caliber 9 mm kurz were produced for the German air force and the Persian police – see there.

Caliber .22 l.r.: Early PPs/PPKs in this caliber have slides, which due to weight reasons, are milled out inside to make them lighter. The recoil would otherwise not be strong enough to self-load the blowback action. So with later PPs/PPKs the outside of the slide is "slimmed" at the front. The recoil spring with PPs/PPKs in caliber .22 l.r. is weaker than the one in other calibers.

The signal pin ("loaded chamber indicator") has to be omitted, as it would fire the rim fire cartridge.

The firing pin is shaped differently compared to the ones in the other calibers in order to reach the primer of the .22 l.r. cartridge.

As there is no signal pin, the underside of the slide does not have to be milled through to give access to this pin. So this underside remains massive as a cover for the firing pin – cf. Walther PPs for the Danish police from 1988 (see there).

The extractor is shaped differently as well and has a more contoured "hook".

The magazines for the .22 l.r. cartridges have no inscription on the body and no button in the viewing slot as a loading device.

PPs/PPKs in this small bore caliber were mostly used for shooting practice (e.g. by "Deutsche Reichsbank", the military or police forces).

The US-American author John Pearson

Illustration 2: *PP/PPK frames – "closed" form with asymmetrical sear on left hand side and "open" form with symmetrical sear on right hand side (according to E.J. Hoffschmidt – see bibliography)*

Distribution of PP/PPK-production (about 540 000 PPs and 290 000 PPKs) amongst the different calibers and years of production (estimates only):

7.65 mm Browning:	**about 93%**		
PP:	750 001 – 397 000P	about 502 200 pistols	(1929 – 1945)
PPK:	754 900 – 431 097K	about 269 700 pistols	(1931 – 1945?)
9 mm kurz:	**about 3.5%**		
PP:	782 304 – 321 408P	about 18 900 pistols	(1933 – 1944)
PPK:	795 900 – 303 587K	about 10 150 pistols	(1933 – 1941)
.22 l.r.:	**about 3.5%**		
PP:	776 100 – 314 000P	about 18 900 pistols	(1932 – 1944)
PPK:	778 500 – 430 960K	about 10 150 pistols	(1932 – 1945?)
6.35 mm Browning:	**about 0.075%**		
PP:	762 590 – 910 941	about 405 pistols	(1932 – 1936)
PPK:	774 660 – 930 650	about 220 pistols	(1932 – 1936)

supports the theory that such guns were used in German concentration camps for executions by a shot in the neck. This theory still needs proof or evidence, however.

Caliber 6.35 mm Browing (.25 ACP): PPs/PPKs in this caliber, too, have slides, which due to weight reasons, are milled out inside to make them lighter.

A Walther Mod. 8 magazine is inserted into the PP/PPK magazine body. Thus the magazine is double-walled and shorter inside.

PPs/PPKs in this caliber are extremely rare, as they were produced in very limited numbers only. PPs and PPKs in the more powerful calibers 7.65 mm and 9 mm kurz were only slightly more expensive and hence more popular. So presumably the production of PPs/PPKs in this caliber started in 1932 and ended in 1936 already.

Oddly enough the some Swiss "Kanton" and municipal police forces bought a number of these weapons as service pistols in the 1930s.

Design features of Walther PP/PPK pistols with Dural frames:

Duraluminium or short Dural is a lightweight alloy material, which was produced from 1909 by "Dürener Metallwerke". It is a wrought alloy consisting of aluminum, copper and some magnesium (Al-Cu-Mg) – a cold-hardening, high-strength, corrosion- and aging-resistant material, which cannot be salt blued like steel though. So early Dural frames were anodized in a dark blue tone. Later this process was replaced by a coating with black lacquer, as this method is simpler and cheaper. Towards the end of World War Two some Dural PPs/PPKs were sold with frames still in-the-white or coated with a clear varnish.

Both PP and PPK pistols were produced with Dural frames, but not especially designated – cf. post-1945 Walther PPK-Ls (see there). There were no Dural slides produced.

Unlike post-1945 Dural PPs and PPKs the ones produced in Zella-Mehlis came in the powerful caliber 9 mm kurz as well.

From 1933 to about 1939 the Dural PP/PPK pistols (e.g. the ones for the "Polizeidirektion München") usually have the so-called "open", non-reinforced frame with a symmetrical sear. From 1939 the Dural frames are not milled through below the sear ("closed" form) and therefore they have an asymmetrical sear, as the latter has to be inserted from the side and not from the underside – cf. illustration 2 and PPs/PPKs in caliber 9 mm kurz.

The barrels for Dural frames were first secured by a pin only. Later a collar was added to the rear of the barrel (at the chamber) to make sure that the barrel could not "creep" to the front. This collar was first in a narrow, later in a wider shape.

It is apparent that the Dural frames did not undergo the same technical changes as the steel frames during their production period. So they kept the short frame tang at the rear and mostly the milled step at the trigger guard base (cf. PPs/PPKs of the 5th and 6th variation). It may be assumed that Dural frames were produced up to 1943 only. Later Dural PPs and PPKs were obviously produced with frames from old stock.

Dural PP:	802 229 - 303 222P
Dural PPK:	802 944 - 398 389K and 417 158K - 431 097K

1. Walther PP and PPK Zella-Mehlis for the German "Wehrmacht" (military forces)

Officers and high-ranking "Wehrmacht" officials were so-called "Selbsteinkleider" (i.e. staff who had to clothe and equip themselves out of their own pocket). So they had to buy their own service pistol (often a Walther PP/PPK or a Mauser HSc resp. a Sauer 38H) including a brown belt holster. They were paid a grant for these expenses.

Up to September 1939 (i.e. the beginning of World War Two) they usually bought their service pistols on the commercial market. Then those pistols have no "Waffenamt" acceptance stamps and only commercial proof marks.

From September 1939 onwards these purchases were done mostly via the ordnance depots. So these pistols have "Waffenamt" acceptance stamps and commercial proof marks as well. The price in those days was 31.20 Reichsmark.

One of the reasons why officers no longer purchased or were able to purchase their service pistols on the commercial market was, that many officers were commissioned while they were on the front lines. So they could not really wait with their purchase until they were on home leave again.

Furthermore there was a sales stop for personal defense pistols. In a letter dated 16th May 1940 Walther announced that in a few weeks they would be able again to deliver PP/PPK pistols, at least to a limited extent, to "Wehrmacht", police, factory security and air raid shelter staff.

These privately purchased pistols served as personal defense weapons mainly in the spare time and on home leave. They were, however, registered in the pay book and included in the inventory of the unit.

Next to their service pistol many officers bought a smaller pocket pistol as a private weapon. Those were, however, normally not used as an official service side arm, but as a secondary weapon (so-called "backup" guns). This was not quite in the sense of the military service regulations though.

For their deployment at the front army officers were issued – if available in sufficient quantities – a pistol in the caliber 9 mm Luger (e.g. a Luger P08 or a Walther P.38) as a so-called "battle pistol". Other military ranks, including non-commissioned officers, were issued a 9 mm Luger pistol as their standard side arm – if available. They received these from the ordnance depot of their unit. If their duty demanded the use of a smaller side arm (e.g. when working in restricted space) these military ranks received e.g. a Walther PP/PPK and a belt holster from their ordnance depot. Those Walther PPs/PPKs were then acquired by the "Wehrmacht" via the army ordnance office ("Heereswaffenamt"). So next to the commercial proof marks they have the "Waffenamt" acceptance stamps – first "Eagle over 359", later "Eagle over WaA 359" (see appendix).

The army ordnance office acquired arms for the army, the navy and the air force, but also for the "Waffen SS", other authorities and sometimes even for allies.

On 20th October 1941 – according to the invoice – Walther sent a PPK with SN 270 240K to the air force for tests. The army had already acquired Walther PP and PPK pistols before this date.

The standard military caliber of Walther PPs/PPKs for the German military was caliber 7.65 mm Browning (.32 ACP). The air force, however, acquired a number of PPs in caliber 9 mm kurz. The standard air force belt holster was either the rather costly "Theuermann" holster or other less costly ones, like the "Anuschat" belt holster.

The PP/PPK pistols delivered to the "Wehrmacht" came with two magazines each – without extensions.

By order of "Oberstes Heereskommando" (superior command of the armed forces), dated 8th November 1943, the carrying of pistols in caliber 6.35 mm or even smaller was forbidden. Carrying such small bore side arms was still quite popular even afterwards – although not necessarily at the front line.

Serial number ranges and markings of "Wehrmacht" PPs/PPKs:

- **PP (caliber 7.65 mm Browning - .32 ACP)**. SN 163 500P – 168 191P: "Eagle/359" on left hand side of slide and left hand side of frame tang
- **PP (caliber 7.65 mm Browning - .32 ACP)**. SN 188 655P – about 383 000P: "Eagle/WaA 359" on left hand side of slide and grip (in front of the grip panel), some PPs with numbered magazines (SN along the length of the floor plate) – air force only?
- **PP (caliber 9 mm kurz - .380 ACP)**. SN 198 359P – 199 833P and 201 979P – 202 476P: "Eagle/WaA 359" (see above), mostly with numbered magazines (see above)
- **PP (both calibers)**. From about SN 383 000P: mostly without "Waffenamt" acceptance stamps and proof marks; these PPs were presumably no longer issued to the "Wehrmacht", but were on stock when the US army occupied the factory; complete pistols or assembled for US soldiers from parts on stock – so-called "GI assemblies" (often with non-matching slides)
- **PPK (caliber 7.65 mm Browning - .32 ACP or very rarely: caliber 9 mm kurz - .380 ACP)**. SN 270 210K – 276 645K: "Eagle/359" on left hand side of slide and left hand side of frame tang
- **PPK (caliber 7.65 mm Browning - .32 ACP or very rarely: caliber 9 mm kurz - .380 ACP)**. SN 298 812K – 424 619K: "Eagle/WaA 359" on left hand side of slide and grip (in front of the grip panel), some PPKs with numbered magazines (SN along the length of the floor plate) – air force only? (cf. picture 20)

Remarks:

It is questionable whether additional stamps – e.g. "Eagle/2" for the air force or swastika (with "M" in a circle) for the navy – are authentic. Such stamps were often used by fakers to "enhance" any pistol.

2. Walther PP and PPK Zella-Mehlis for Government Authorities and other Institutions:

a) Police:

Police pistols can only rarely be assigned to a special police force or unit, as they were not marked consistently. If at all, police markings were mostly stamped on the front gripstrap, resp. on the backstrap or the frame. By decree of Heinrich Himmler the marking of police weapons should have been omitted from mid-1936. This decree was not implemented by all police forces though.

The Ministry of the Interior was in charge of procuring all police weapons.

The following police markings are known. Some of them cannot be assigned clearly to certain police authorities or other institutions. Possibly some of these markings are not police stamps at all, but used by factories or factory securities.

- "**P.V.Gö.**" (plus inventory number): "Polizeiverwaltung ...." – police administration)?
- "**P.V.E.**" (plus inventory number) – police administration?
- "**P.D.S.**" (plus inventory number)
- "**P.M.P.**"
- "**K.H.**" (plus inventory number)
- "**N.D.L.**" (plus inventory number)
- "**BR.P.P.K.N:r**" (plus inventory number)
- "**P**" (in a circle, plus inventory number): PPKs only in the SN range 771 194 – 778 462
- "**P.M. 1940**" (on the frame, above the SN): PP SN range 174 686P – 176 130P and 269 461P – 269 500P; PPK SN range 275 252K – 275 528K and 349 080K – 355 592K (possibly "Polizeischule Münster/Hiltrup – a police academy)
- "**P.M.P.**": Polizei Mecklenburg-Pommern ?
- "**ST.P.S.**" (plus inventory number): PPK SN range 851 728 – 852 170 and 901 054 (presumably "Staatspolizei Saarland" = "Gestapo")
- "**O.L.B.**" (on right hand side of frame)
- "**C/R**" (in a circle)
- "**L.S.K.**" (plus inventory number)
- "**B.Wi. 1/1310**"
- "**Rz 418**" (on left hand side of frame)
- "**W/M/B**" (as a triangle on frame or trigger guard)
- "**EHR**" (on frame and slide)

Other police markings can be assigned clearly to certain police authorities:

"PDM" (plus inventory number) vertically on left hand side of frame for "Polizeidirektion München":

PP	811 004 - 812 635 (closed SN block)
PP	885 866 - 885 906 (closed SN block)
PP	941 741 - 941 845 (closed SN block)
PPK	852 571 - 853 425 (no closed SN block, no "PDM" markings, but other typical "PDM" characteristics - see below)
PPK	899 001 - 899 515 (closed SN block)
PPK	922 648 - 922 763 (closed SN block, no "PDM" markings, but other typical "PDM characteristics - see below)

Although some of these "PDM" pistols are not marked with the "PDM" stamp, they can clearly be assigned to "PDM", as they show all the typical other characteristics: bottom magazine release (although not in caliber 9 mm kurz), barrel with a bulge at the muzzle, PP frames made of salt blued steel, PPK frames made of anodized Dural.

The PP/PPK pistols delivered to the "PDM" came with two magazines each – with extensions.

From 1939/40 PPs/PPKs, which were issued to the police, were stamped with a police acceptance stamp – first "Eagle/C", then "Eagle/F":

PP	205 670P - 348 094P: "Eagle/C"
PP	354 444P - 379 433P: "Eagle/F"
PPK	1 002 794 - 413 086K: "Eagle/C"
PPK	423 537K - 430 090K: "Eagle/F" (but: 428 063K - 428 288K: "Eagle/C" or "Eagle/F")

Remarks:

PPs/PPKs with the police acceptance

stamp "Eagle/L" are presumably not authentic. This stamp was normally used for other pistol models.

PPs/PPKs with both "Waffenamt" and police acceptance stamps must be regarded critically, too.

b) "Reichssicherheitshauptamt" ("RSHA" resp. SS):

The "Reichssicherheitshauptamt" was established by "Reichsführer SS", Heinrich Himmler, on 27th September 1939 by a merging of "Sicherheitspolizei" (i.e. "Kripo" and "Gestapo") and "Sicherheitsdienst". Then it was one of the twelve "Hauptämter" (main authorities) of the SS.

This institution is dealt with separately here, although it was a sort of police authority. The weapons for the RSHA/SS – like the ones for the police – were procured by the Ministry of the Interior.

Weapons, which were in the possession of the SS before 1938, are listed here, too, as they were taken over when the RSHA was established.

The following classification of the RSHA/SS PPs and PPKs is based on Steven J. Stepan, "SS- Walther PP/PPK Identification & Documents, 2010. This booklet is a revised and translated edition of J. Görtz & D. H. Marschall, "SS Walther PP/PPK 1939 – 1944", 1999.

One question still needs clarification: Why were "RZM" marked "Type 1 RSHA PPKs" used by the RSHA, although as a rule the "Reichszeugmeisterei" ("RZM") procured equipment exclusively for the NSDAP ("Nazi Party")?

The serial numbers given below are mostly based on the author's database and are perhaps incomplete.

Walther PP type 1 RSHA:

Characteristics:

SN on frame and additionally stamped on slide (in a semi-circle below the muzzle); magazines numbered on spine.

SN range: The author is not aware of any reference specimens.

Walther PPK type 1 RSHA:

Characteristics:

"RZM" emblem on left hand side of slide; SN on frame and additionally stamped on slide (in a semi-circle below the muzzle); some magazines numbered on spine.

SN range: 839 326 – 841 729

Walther PP type 2 RSHA:

Characteristics:

SN on frame and additionally stamped on right hand side of slide (without suffix "P"); magazines numbered on spine (without suffix "P").

SN range: 977 904 – 982 813

Walther PPK type 2 RSHA:

Characteristics:

no "RZM" emblem on left hand side of slide; SN on frame and additionally stamped on slide (in a semi-circle below the muzzle); some magazines numbered on spine.

SN range: 845 483 – 1 006 697 and 192 067K – 192 925K

Walther PP type 3 RSHA:

Characteristics:

SN on frame and additionally stamped

on right hand side of slide (with suffix "P"); magazines numbered on spine (with suffix "P"). In all three locations the suffix is stamped below the serial number.

SN range: 124 774/P - 125 168/P

Walther PPK type 3 RSHA:

Characteristics:

SN on frame and additionally stamped on right hand side of slide (with suffix "K"); magazines numbered on spine (with suffix "K"). In all three locations the suffix is stamped below the serial number.

SN range: 206 019/K - 234 832/K

Walther PP type 4 RSHA:

Characteristics:

SN on frame and additionally stamped on right hand side of slide (with suffix "P"); magazines numbered on spine (with suffix "P"). In all three locations the suffix is stamped in the standard location behind the serial number.

SN range: 144 179P - 223 174P

Walther PPK type 4 RSHA:

Characteristics:

SN on frame and additionally stamped on right hand side of slide (with suffix "K"); magazines numbered on spine (with suffix "K"). On frame and slide the suffix is stamped in the standard location behind the serial number, on the spine it is stamped below the serial number though.

SN range: 242 844K - 334 500K

Walther PP type 5 RSHA:

Characteristics:

This type 5 was obviously not produced, unless pieces of evidence appear in the future.

Walther PPK type 5 RSHA:

Characteristics:

SN on frame and additionally stamped on right hand side of slide (with suffix "K"); magazines numbered on floor plate (along its length, with suffix "K"). In all three locations the suffix is stamped behind the serial number. These magazines should not be confused with the ones of the "Wehrmacht". The latter had the SN on the floor plate, too – but an additional "Waffenamt" acceptance stamp.

SN range: 342 144K - 426 654K

Delivery of Walther PPKs to the "Sicherheitspolizei" ("Sipo") and the "Sicherheitsdienst" ("SD") of the "Einsatzgruppe C. Lemberg" (task force in Lemberg/Western Ukraine, today named Lwiw) in July 1941:

PPK	296 004K - 296 999K (almost closed SN block with the exception of a few PPKs, which were issued to other RSHA departments)
PPK	302 001K - 302 203K (closed SN block)
PPK	306 442K - 307294K (almost closed SN block with the exception of a few PPKs, which were issued to other RSHA departments)

c) "Reichsbahndirektion" ("RBD") railway police:

Many PPs/PPKs were issued to the railway police of the different "Reichsbahndirektion" "("RBD") authorities. These RBD pistols can be identified by one or more of the following features: SN subsequently stamped on the left hand side of the magazine and/or on the right hand side of the slide; date "1940" stamped additionally on the left hand side of the slide; RBD property mark on the front gripstrap, belt holster marked with property stamp.

The following RBD PPs and PPKs are known so far:

PP 982 794: two numbered **mags., holster marked** "RBD Stuttgart 1937"
PP 983 098: numbered mag., holster marked "RBD Stuttgart 1940", SN on right hand side of slide
PP 991 720 - 992 741: gripstrap marked "RBD Münster (Westf.)"
PP 992 707 - 992 741 with holster marked "RBD Münster (Westf.)" (closed SN block?)
PP 993 472 - 993 524: gripstrap marked "RBD Münster (Westf.)" (closed SN block?)
PP 122 942P: SN on right hand side of slide, numbered mag.
PP 130 231P: SN on right hand side of slide, numbered mag., "1940" on slide
PP 138 859P: SN on right hand side of slide, numbered mag.
PP 148 280P: "1940" on slide
PP 148 808P: SN on right hand side of slide, "1940" on slide
PP 153 767P - 153 770P: gripstrap marked "RBD Münster (Westf.)" (closed SN block?)
PP 157 535P: holster marked "RBD Halle"
PP 157 705P: holster marked "RBD Halle"
PP 158 581P: "1940" on slide holster marked "RBD Stuttgart 1940",
PP 158 630P: "1940" on slide, holster marked "RBD Stuttgart 1940", numbered mag.
PP 158 632P: "1940" on slide, holster marked "RBD Stuttgart 1940", numbered mag.
PP 160 025P: holster marked "RBD Halle", but commercial pistol?
PP 160 220P - 160 248P: "1940" on slide, SN on right hand side of slide, numbered mag. (closed SN block?)
PP 179 829P: "1940" on slide
PP 179 902P: "1940" on slide, SN on right hand side of slide, holster marked "RBD Stuttgart 1940", two numbered mags
PP 179 962P: "1940" on slide, SN on right hand side of slide, holster marked "RBD Stuttgart 1937", numbered mag.
PP 181 434P: "1940" on slide, SN on right hand side of slide, numbered mag.
PP 181 625P: "1940" on slide, SN on right hand side of slide
PP 209 031P: "1941" on slide, SN on right hand side of slide
PPK 172 837K - 172 839K and 290 686K - 290 688K: "RBD Münster (Westf.)" on front gripstrap

d) "Deutsche Reichspost" ("DRP"):

The postal authorities procured Walther PPKs only, no PPs, for the armed guards of the postal value transports.

PPK 790 639 - 913 671 (no closed SN block; "D.R.P." written with periods, vertically on left hand side of frame)
PPK 953 866 - 955 817 and 955 900 - 959 809 (no closed SN block; "DRP" written without periods, vertically on left hand side of frame)

Some DRP PPKs have a nickel-plated finish (applied subsequently?).

Uniformed postal officials carried their PPK in a black AKAH belt holster, which was actually made for the slightly bigger PP. The insides are stamped with "DRP" and the abbreviation for the postal district.

e) "Reichsfinanzverwaltung" ("R.F.V."):

The "Reichsfinanzverwaltung" was, amongst other tasks, in charge of customs. Their officials were armed with Walther PPs and PPKs, but with other types of pistols from other manufacturers as well.

The Walther PP was usually carried by uniformed customs officials, whereas the plain clothes ones (e.g. from the customs investigation service) were issued PPKs.

PP 765 122 - 316 616P (no closed SN block, "R.F.V." stamp with suffix "W" and inventory number on backstrap or on front of grip)

PPK 774 836 - 327 622K (no closed SN block, "R.F.V." stamp with suffix "KW" and inventory number on front of grip)

The meaning of the suffixes "W" and "KW" is unknown. The assumption that they stand for "Waffe" (gun) and "Kurzwaffe" (hand gun) is only a theory without any proof.

Interestingly enough some "R.F.V." PPs were used by the Danish state police ("Rigspolitiet") in post-1945 days – see there.

f) "Reichsjustiz" ("R.J."):

The "Reichsjustiz" procured Walther PPs for the judicial police in courtrooms and for prison guards. These pistols are marked "R.J." (early PPs on the front of the grip, later PPs on the backstrap).

The interpretation that "R.J." stands for "Reichsjugend" (Hitler Youth) is not correct.

Some PPs marked "R.J." were used in post-1945 days by the judicial service of the Federal Republic of Germany.

PP 818 675 - 331 846P (no closed SN block)

g) "Deutsche Reichsbank" ("RB") and Commerz- und Privatbank ("C.P.B"):

The "Deutsche Reichsbank" procured Walther PP/PPK pistols for the protection of cash transports and also bank counter staff.

A special shoulder holster was issued for the concealed carrying of these pistols.

The bank counter staff kept them in open, especially designed, wooden boxes below the counter. For storage the "Reichsbank" used special boxes, which were not identical with the normal Walther boxes. Those boxes were of an anthracite color outside and lined with green velvet inside. On the outside and on the inside there were various information and warning labels for the bank staff.

The "Reichsbank" maintained several armories, which serviced and proof shot those pistols on a regular basis. Every service and proof shooting was entered into a special file, which was kept with the pistol.

The "Reichsbank" also issued its own service regulations for the Walther PPs/PPKs.

The standard caliber for the "Reichsbank" personal defense pistols was 7.65 mm Browning. For shooting practice the "Reichsbank" also acquired a number of PPs/PPKs in caliber .22 l.r. These small bore pistols were usually handed out with magazines with green magazine extensions to make sure they could be differentiated.

All "Reichsbank" PPs/PPKs have the serial number vertically on the grip and additionally vertically on the right hand side of the slide – hand stamped or pantographed (right in front of the slide serrations). Early "Reichsbank" PPs/PPKs carry the serial number additionally on the side of the magazine extension.

Both the front and the rear sight are usually painted with a contrasting color.

It is evident that especially early PPs/PPKs were "upgraded" by fitting the more solid one-piece firing pin with the "slim" safety lever instead of the fragile two-piece firing pin together with the "fat" safety lever.

PP 750 056 - 990 476 (no closed SN block)

PPK 763 924 - 343 957P (no closed SN block)

One more major bank procured Walther PPKs for its staff: the "Privat- und Commerzbank" (later named "Commerzbank" only), which was founded in 1920.

The Walther PPKs for this bank are marked on the right hand side of the slide with "C.P.B.".

This abbreviation was first interpreted as a short form of "Criminal Polizei Berlin", but new research efforts – especially by Joseph Wotka/USA – have shown that the other interpretation is more likely.

PPK 764 194 - 771 909 (no closed SN block)

h) "Reichsrundfunkzentrale" ("R.R.Z."):

There are only very few Walther PPs known, which carry on the right hand side of the slide the engraved designation "R.R.Z." (plus inventory number). One theory states that these PPs were issued to the "Reichsrundfunkzentrale" (radio station head office) in Berlin. This theory still needs proof or evidence, however.

i) "Standschützenverband Tirol-Vorarlberg" ("T.ST.V."):

After the annexation of Austria by the German Third Reich in March 1938 the traditional shooting association ("Tiroler Standschützen"), which was already founded in the 16th century, was merged into the conformist, paramilitary association "Standschützenverband Tirol-Vorarlberg". This was done by the "Nazi Gauleiter" Hofer from "Gau Tirol-Vorarlberg".

From 1938 to 1944 there were regular shooting competitions – first with "Mauser Wehrmanngewehr" (caliber 8,15 x 46 R), from 1941 onwards with Walther PPKs in caliber .22 l.r. and with small bore rifles.

However, there is no Walther PPK in caliber .22 l.r. known with the Tyrol marking. Only the "Wehrmannsgewehre" and some Walther PPs in caliber 7.65 mm Browning show this marking: "Tyrol Eagle" (with swastika and oak leaves). So it is questionable whether these "Tyrol PPs" were awarded as shooting prizes or issued as defense weapons to officials of the "T.ST.V.". The fact that the purchase of these pistols was already done in 1938 leads to the conclusion that the latter possibility is more probable.

The Tyrol PPs are in the same serial number range as the ones for the "Gruppen der SA der NSDAP" – see below.

PP 135 315P - 137 352P (closed SN block?)
PP 137 568P
PP 138 143P - 138 172P (closed SN block?)
PP 139 935P
PP 139 172P
PP 139 247P - 139 266P (closed SN block?)

3. Walther PP and PPK Zella-Mehlis for "Nationalsozialistische Deutsche Arbeiterpartei" ("NSDAP"):

a) Groups of "SA der NSDAP":

Not all the regional and other groups of the "SA der NSDAP" obviously marked their pistols. After all, only the markings of the groups mentioned below are known. There are no closed SN blocks, but only certain batches of pistols distinguishable, which were issued to the same SA group.

These SA groups acquired about 4850 Walther PPs in caliber 7.65 mm Browning. PPs in caliber .22 l.r. and PPKs were procured only very rarely. The SA pistols were normally marked on the front gripstrap (rarely on the slide) as follows below:

"**SA der NSDAP**... (with the following additional designations)

Picture 21: *Walther PP ZM, 4th variation (1938), SN 138 935P, caliber 7.65 mm Browning, (.32 ACP), "Tiroler Standschützen" emblem on left hand side of frame*

...Standarte Feldherrnhalle"
...Standarte 92"
...Stab der SA Gruppe Westfalen" (e.g. on an engraved and gold-plated PPK!)
...Oberste Führung" (authentic?)
...Gruppe Alpenland"
...Gruppe Berlin-Brandenburg"
...Gruppe Franken"
...Gruppe Hochland"
...Gruppe Hansa"
...Gruppe Hessen"
...Gruppe Hochland"
...Gruppe Niederrhein"
...Gruppe Niedersachsen"
...Gruppe Nordmark"
...Gruppe Mitte"
...Gruppe Oder"
...Gruppe Ostmark"
...Gruppe Pommern"
...Gruppe Sachsen"
...Gruppe Schlesien"
...Gruppe Südmark"
...Gruppe Südwest"
...Gruppe Thüringen"
...Gruppe Westfalen"

The members of the SA target shooting teams used Walther PPs with reddish brown grips. Each SA shooting team member was given one pair of reddish brown grips and each year an especially selected PP by Walther. The reddish brown grips were then changed over to the new PP each year.

The SA PPs were carried in coffee-brown leather belt holsters model "Anuschat" made by the company Albrecht Kind ("AKAH"). On the flap these holsters have the embossed SA emblem. AKAH produced these holsters exclusively for the SA at a price of 4.50 Reichsmark – about one tenth of the price of a Walther PP (46.50 Reichsmark). For a comparison: The wage of a factory worker in those days was about 200.00 Reichsmark per month.

Picture 22: *Walther PPK ZM, 3rd variation (1935), SN 835 476, caliber 7.65 mm Browning (.32 ACP), "RZM" emblem on left hand side of slide*

b) "Reichszeugmeisterei" ("RZM"):

The RZM was in charge of the procurement of weapons and equipment for all the bodies of the NSDAP party. They purchased about 60 000 PPKs and PPs (almost exclusively PPKs though). According to the handbook of the NSDAP the Walther PPK was the standard service pistol of that party from 1934 onwards.

As a rule all equipment, which was procured by RZM, was marked with their emblem ("RZM" in a circle). The Walther pistols carried that emblem on the left hand side of the slide. To make space for this emblem the whole slide inscription had to be moved towards the muzzle. Some faked RZM pistols can be recognized as fakes, if this is not the case. Unfortunately fakers learn quickly.

The RZM belt holsters were made by the company Albrecht Kind ("AKAH"), too. These belt holsters model "Anuschat" were made of havanna-brown leather with orange-red seam material and a closing strap, which was cut off acutely angled at the front. On the side flap there is the AKAH logo and "D.R.G.M" ("Deutsche Reichs-Gebrauchsmuster" – a minor patent) next to an embossed "R.Z.M. No. 1051".

PP	882 725, 218 496P, 253 885P, 254 355P, 265 542P, 278 439P, 290 076P (Only these specimens are known. It is questionable whether they are authentic or not)
PPK	803 258 - 821 490 (no closed SN block)
PPK	821 500 - 844 470 (closed SN block)
PPK	844 470 - 289 000K (no closed SN block)
PPK	289 089K - 289 293K (closed SN block)
PPK	290 000K - 372 198K (no closed SN block)

c) Honor Weapons for Political Leaders ("PL"):

For the political leaders ("Parteileiter") of the NSDAP party, too, the Walther PPK was the standard service pistol. It should first be mentioned here that this variation of the Walther PPK is perhaps the most often faked one.

The term "Ehrenwaffe" (honor weapon) suggests that these pistols were honorary awards – quite like medals or badges of honor. This is not the case, however. The PLs had to purchase their honor weapons themselves and pay for them out of their own pockets. This purchase was done e.g. via the RZM – sometimes paid in instalments. So it was only an honor to be allowed to carry such a PL pistol. Hence there were no certificates of award, but only permits to buy and carry these pistols. These permits were issued to the PLs by their superior PL after they had done some shooting practice. According to the NSDAP handbook every PL was obliged to fulfil certain conditions on the shooting range every year.

These PPKs were carried by the PLs together with the golden-brown uniform. Up to 1938 this was done in a standard havanna-brown RZM belt holster.

From 1938 the company Albrecht Kind provided a special model "Anuschat" holster again made of havanna-brown leather with orange-red seam material. There is a big eagle with a swastika ("Reichsadler") embossed on the flap. To make sure that this emblem is not partially hidden, the closing strap is cut off in a right angle.

From 1939 the "Reichsadler" was no longer embossed on the leather of the flap, but replaced by one made of gold-colored metal, which was mounted on the flap. This was certainly more prestigious. One of the first ones to carry this new type of holster was of course Dr. Robert Ley NSDAP "Reichsorganisationsleiter", who had good reasons to make a good impression at that time.

AKAH produced these belt holsters exclusively for the PLs at a price of 4.35 Reichsmark – about one tenth of the price of a Walther PPK. For a comparison: The wage of a factory worker in those days was about 200.00 Reichsmark per month.

Because of their golden-brown uniform and their "flashy" outfit the PL were often called "gold pheasants" by commoners.

The one-piece PPK grips belonging to the PL PPKs display a "Reichsadler" (eagle with swastika) on both sides. Walther could verifiably provide these grips only from spring 1938 onwards. On 29th January 1938 Walther sent a sample to the NSDAP headquarters for assessment. On 9th February 1938 the headquarters decreed that in future all PL PPKs had to be supplied with this type of PL grips. This was implemented by Walther 14 days later.

As a rule PL grips were mottled brown. From 1944/45 onwards they obviously came in black, too.

There are three variations of PL-PPKs:

Type 1: PPK without "RZM" emblem, without PL grips, in a PL belt holster
Type 2: PPK with "RZM" emblem, without PL grips, in a PL belt holster
Type 3: PPK without "RZM" emblem, with PL grips, in a PL belt holster
(cf. picture 19)

So fakes or forgeries are easily made. If any PPK before SN 172 000K has PL grips, then these grips can be fakes. Or they can be authentic PL grips mounted on an early PPK. It was of course essential for any PL to "upgrade" his pre-1938 PPK with the more prestigious PL grips and be seen with it.

Unfortunately there are very well-made faked PL grips on the market in the meantime. Thus it is quite easy to "upgrade" any PPK to a PL PPK.

The same is true for PL belt holsters: If

they show the Havana brown leather, orange-red seam material, the "Reichsadler" and the closing strap, which is cut off in a right angle, they can always be original or well-made fakes.

d) "Nationalsozialistisches Kraftfahrerkorps" ("NSKK"):

This organization arose out of the "Nationalsozialistischer Automobilkorps" and was renamed in 1931 to NSKK. In August 1934 the "Motor-SA" and the "NSKK" were merged and placed directly under A. Hitler's command.

From 1934 the NSKK was in charge of the road safety education and the training of drivers – including the youth.

A great number of the car and motorbike racing drivers of that era were members of the NSKK.

Only high-ranking leaders of the NSKK carried a side arm with their uniform. As a rule this was a Walther PP and only rarely a PPK.

These NSKK pistols were marked with an NSKK emblem on the left hand side of the slide ("NSKK" over a "Reichsadler" with a swastika). To make space for this emblem the whole slide legend had to be moved towards the muzzle – cf. picture 16, too.

The NSKK PPs were carried with the uniform in black leather belt holsters model "Anuschat" made by the company Albrecht Kind ("AKAH"). On the flap these holsters have the embossed NSKK emblem. As with PL holsters the closing strap is cut off at a right angle. AKAH produced these belt holsters exclusively for the NSKK at a price of 4.50 Reichsmark – about one tenth of the price of a Walther PP (46.50 Reichsmark).

PP	810 202 - 200 737P (no closed SN block)
PPK	807 235 - 862 658 (no closed SN block), 198 278K, 297 336K.

Remarks:

Walther PPs and PPKs were quite popular as shooting prizes with NSDAP organizations, too. Mostly those pistols were engraved with corresponding inscriptions on the slide.

4) Walther PP/PPK for Non-German Armies and Police Forces:

a) Danish Police Forces ("Rigspolitiet" – "Rplt."):

The Danish police procured Walther PPs/PPKs before 1939 and carried on doing so after Denmark was occupied by German troops in April 1940, as this occupation was officially seen as a "friendly takeover". The Danish police bought some Browning model 10/22 pistols, but mainly Walther PPs and a few PPKs (a total of about 1350 pistols) – all in caliber 7.65 mm Browning. Some 9 mm kurz Walther PPs, which were in storage in Danish arsenals, were rebuilt for the caliber 7.65 mm Browning. These PPs kept their bottom magazine release though.

First the PP/PPK pistols for the Danish police were not marked with any stamps. Only from about SN 182 226P onwards did the police ordnance staff mark them (on the left hand side of the frame in front of the trigger axle) with "Rplt." (plus inventory number). This was not done consistently, however. So there are some pistols, especially PPKs, known without that marking.

As some pistols were confiscated as "war booty" after World War Two and were given to the "Rigspolitiet", there are all sorts of combinations of "Rplt." and other stamps possible ("Rplt." with "RZM", "R.F.V.", "WaA 359" etc.).

When "Rplt." PPs and PPKs were refurbished in Danish arsenals, some of them were given a new finish – e.g. frames and/or slides were matte chrome-plated.

Most "Rplt." PPs have a swiveling lanyard ring at the bottom of the grip.

Picture 23: *Walther PP ZM, 4th variation (1938), SN 120 025P, caliber 9 mm kurz (.380 ACP), Persian contract, with Persian slide legend on left hand side of slide*

PP	127 732P - 150 948P (no closed SN block, mostly without "Rplt." stamp)
PP	182 226P - 366 329P (no closed SN block,mostly with "Rplt." stamp and inventory number)
PPK	947 846 - 354 293K (no closed SN block, mostly wi thout "Rplt." stamp)

b) Norwegian Police Forces ("Politi"):

The Norwegian police, too, procured Walther PP and PPK pistols before and during the occupation by German troops. There were, however, no purchases "en bloc". So there are no "closed" SN blocks.

The Norwegian police PPs and PPKs were marked on the right hand side of the slide with "Politi" (plus Norwegian coat of arms and inventory number) – cf. Norwegian P38s.

c) Persian Police and Military Contract Pistols (PPs only):

Between 1939 and 1940 Walther made three deliveries of PPs (a total of about 9050 pistols) to the regime of the Shah of Persia, Reza Pahlewi, in order to arm the police. Many of these pistols were used by the Persian army later, however.

All of the Persian PPs are in caliber 9 mm kurz (.380 ACP) and have the bottom magazine release. The slide inscription was moved from the left hand side to the right hand side of the slide. The left hand side of the slide shows the Persian coat of arms in the middle. On its left there is written in Persian Farsi "1317" (according to the Arabian calendar, which corresponds with our year 1939). On the right hand side of the coat of arms is written "State Police" in Farsi. On the left hand side of the frame there is the arsenal number in Farsi. There are German and sometimes Persian acceptance and proof stamps in different places.

PP 116 118P - 120 919P (closed SN block)

PP 152 867P - 152 967P (no closed SN block)

PP 169 427P - 173 651P (closed SN block, mostly with an additional SN on right hand side of slide, vertically in front of the slide serrations - similar to the pistols of the "Deutsche Reichsbank"

The following Walther PPs have a differing Persian designation and presumably were part of another Persian contract. In the Walther literature they are sometimes erroneously assigned to a contract with the Arabian state Yemen:

PP 147 463P - 147 649P (no closed SN block): Walther inscription on right hand side of slide, on left hand side of slide - in Farsi (from the left to the right): letter "F" / Persian acceptance stamp (= a stylized shell) / letter "Y"

d) Lithuanian Contract Pistols (PPKs only):

In 1937 Walther shipped about 200 Walther PPKs to Lithuania. These are marked with the "Pillars of Gediminas", a symbol used by the former Grand Duke of Lithuania.

PPK 955 892 - 956 660

e) Polish Contract Pistols (PPKs only):

Shortly before the beginning of World War Two in 1939 Walther shipped a number of PPKs to Poland. These carry the Polish military acceptance stamp ("G over 2" in an oval) on the left hand side of the frame.

PPK 220 830K - 226 518K (closed SN block?)

f) Swedish Police Forces (almost exclusively PPs only):

In World War Two Sweden officially took a neutral position and was not occupied by German troops. But German troops had the right to pass through Swedish territory on the way to and from Norway. Swedish iron ore, which was of a great strategic importance for Germany was constantly shipped to Germany. Amongst other factors this "good conduct" of the Swedish led to the fact that Sweden received Walther pistols before and during World War Two – mostly for the Swedish police.

Most of these pistols were PPs, but some PPKs were delivered, too. The major Swedish importer was a company named J.V. Astroem in Bjurtjaern/Sweden.

As the Swedish police carried their pistols with a long lanyard around the neck, these PPs were delivered by Walther with the standard factory-made, swiveling lanyard ring. PPKs already have a lanyard loop as an integral part of the pistol design.

PP SN 137 109P marked "S.P.I.F." (Stockholm police?)

PP SN 166 608P marked "B.S.NM-217-S" (Swedish police?)

PP SN 167 128P (according to the invoice sold to a Swedish telephone company)

g) Swedish Power Plant Group "Statens Vattenfallsverk":

From 1942 until 1944 the Swedish power plant "Statens Vattenfallsverk" purchased Walther PPs and PPKs with Dural frames for their factory security. These pistols show all the characteristics of the PPs/PPKs of the 6th variation. On the right hand side of the slide they are inscribed with "Statens Vattenfallsverk" (with an inventory number above the inscription).

PP 806 109 and 208 919P (samples or advance deliveries?)
PP 254 339P - 257 108P (no closed SN block)
PP 313 408P - 313 422P (closed SN block)
PPK 376 623K - 376 837K (closed SN block)

h) Bulgarian Police and Military Forces:

In World War Two the Bulgarian kingdom under King Boris III was officially an ally of Nazi Germany – but only half-heartedly. After internal conflicts in Bulgaria the government asked for aid shipments of weapons from Germany. The chief of the German army ordnance depot decreed that from March 1943 a total of 5000 pistols (P.38s, P08s, PPs and PPKs) should be shipped to Bulgaria. In September 1943 Bulgaria asked for another 2000 pistols in caliber 7.65 mm Browning. Furthermore many service pistols were shipped to Bulgaria by the German army for the German troops there. These stayed in Bulgaria after World War Two as war booty.

Most of these pistols were still used long after the end of World War Two by the Bulgarian police and armed forces.

From 1990 onwards the Bulgarian government gradually sold these service pistols, which were mostly heavily used meanwhile, to western trade companies for "hard currency".

Many of the Walther PPs/PPKs, which were shipped to Bulgaria before 1945, have "Waffenamt", "RZM" or even Persian markings.

During their long years of service in Bulgaria most of these PPs/PPKs were modified – e.g. refurbished, refinished (perhaps even chrome- or nickel-plated), engraved, etc.

If the original grips broke, they were mostly replaced by grips made of grey duroplastic material – but without the Walther banner.

PP 188 665P - 291 988P
PPK 345 189K - 385 308K

i) Swiss Police Forces:

In World War Two Switzerland was officially neutral, but many foreign exchange transactions were handled by the Germans with the help of Swiss banks. Furthermore the Swiss and the German industries were dependent on each other, as the gradual and total enclosure of Switzerland by the Germans and its allies led almost automatically to an intensification of the economic ties of both countries. Switzerland was made to give Germany generous credits. In return Switzerland was allowed – amongst other transactions – to purchase Walther pistols for their police forces even long after the beginning of World War Two. The last known PPK, which was shipped to Switzerland as late as 1944, was SN 416 285K. This pistol went the Solothurn Police.

Most of these Swiss police pistols are not marked as such and there are no closed SN blocks known.

Other Markings and Designations:

Retailers:

Up to now the following retailers' designations are known:

- "Waffen Jung Stuttgart"
- "Vandrey Hamburg"
- "Gewehr Center Cottbus"
- "C. Widmer St. Gallen" (Switzerland)
- "FZ" (Flückiger, Zurich/Switzerland)
- "Glaser Zürich", (Switzerland)
- "GB" (Bürgin, Basel/Switzerland)
- "A.F. Stoeger Inc., New York" (USA)
- "Chas. A. Heyer & Co., Nairobi", (Kenia/ East Africa)

Factory Securities:

- "Thiel" (with an anchor): watch factory Thiel, Ruhla/Thuringia
- "Metall u. Eisen": metal factory, Nürnberg
- "G-Werkschutz": Gustloff-Werke, Suhl/ Thuringia
- "C.L. A-G": Carl-Lorenz AG, Berlin-Tempelhof
- "NSU": vehicle factory, Heilbronn-Neckarsulm
- "Barmag": textile factory, Remscheid
- "S.B.V.": Swiss banking association, Basel/ Switzerland
- "R.W.E.": power plant group, Essen/North Rhine-Westfalia
- "Fa. Mannesmann Langschede": steel company, North Rhine-Westfalia

Export:

The following markings can be found on PPs/PPKs for export:

- "Allemagne"
- "Made in Germany"

Other Variations and Peculiarities

Engraved PPs/PPKs:

The term "factory engraving" is often used in describing engraved PPs and PPKs. It is a fact, however, that the engraving department within the Walther factory was very small and dealt mostly with hunting rifles. In the late 1930s only four employees worked here under the supervision of a foreman named Schneider. So most orders as to engraved PPs or PPKs were – as one would say nowadays – "outsourced" to Thuringian engravers in the Zella-Mehlis area – e.g. Zöller, Spörer or Schilling. In those engraving shops there worked of course not only the master engravers, but also other employees. So the quality of the engravings may vary.

The typical engraving patterns of this 1930s to 1940s era were: oak or vine leaves, floral or scroll engravings. Hunting scenes were the exception. Next to the full engravings there were of course partial engravings: dedications, names, dates of prize awards or days of remembrance, etc.

The final finish could be blued, nickel-, chrome-, silver- or gold-plated.

Often these engraved pistols were provided with special grips and sometimes even magazine extensions made of mother-of-pearl, ivory, carved wood – or simply white plastic.

There are evidently "closed SN blocks" of engraved PPs and PPKs, as whole batches of unhardened PPs/PPKs still in-the-white with consecutive serial numbers were set aside for future orders of engraved pistols. These pistols were hardened only after the engraving, then finished and proof-fired. So there might be discrepancies between the date of production and the date of the proof. If proof marks are overrun by the engraving or even removed, this is very suspicious, as all weapons had and still have to have proof marks by law.

As many US soldiers still remained in Germany after the end of World War Two, some of them used this time to have their "war booty" engraved subsequently by German engravers. So it is of course often very hard to distinguish whether an engraving was made before or right after 1945. The true proof of an authentic engraving can only be an invoice or a similar document. It is of course very helpful, if an engraved PP/PPK is within a "closed SN block" with other engraved PP/ PPK pistols.

PP/PPK fluorescent front sights:

There are two types of these front sights with a contrasting yellow, fluorescent paint, which differ only slightly in the shape of the knurled nut, however. These night sights are pushed onto the normal front sight. They engage in a cut at the rear of the front sight. At the front they are fixed by a knurled nut in a blind hole in the base of the front sight – see picture 19.

These night sights were sold for PPs/

PPKs from 1933 onwards. Since that year PPs/PPKs had that corresponding cut in the rear of the front sight. PPs/PPKs from 1942 onwards and those in caliber .22 l.r. do not have this cut. Obviously these fragile accessories did not prove themselves – at least not in times of war.

Today these fluorescent night sights are regarded as very rare. Hence they are rather expensive. Presumably some of these parts still lead a sad and unrecognized existence somewhere in any box or chest today.

The extra charge for a fluorescent night sight in 1933 was 3.75 Reichsmark.

In 1950 Walther/Ulm made new design drawings for this accessory, but it was never produced again.

PP/PPK stainless steel barrels:

These barrels are marked with "Böhler Stahl Antinit", "nichtrostend", "rostfrei" or stainless (often on the box only), e.g. for pistols, which went into export.

The extra costs for a stainless steel barrel in the 1930s was 3.00 Reichsmark.

PP/PPK insert barrels in caliber 4 mm M20 for PPs/PPKs in caliber 7.65 mm Browning or 9 mm kurz:

This field of interest has been a bit neglected so far, but was recently dealt with extensively by **Martin Klesper**. The results of his research are summarized below.

Already before the introduction of the Model PP in the 1930s Walther offered insert barrels. These barrels carry a mark at their muzzle so that after test firing they can always be refitted in the same position again. For a vertical correction Walther either recommended a change of the aiming point or an adaptation of the front sight – e.g. by a piece of cardboard.

The first variation of the insert barrels still had a flat locking nut. After some shots these nuts often came loose, as they could not be fixed when screwed on. This led to changes in the shooting results and thus to complaints of the customers.

The improved variation had a wider locking nut for a better grip. This was only a stopgap solution. But this variation was already shown in the Walther PP manual dated 1934.

A real improvement came later with a slotted cutout at the muzzle with two corresponding cams at the front of the ejector. With these the insert barrel can be fixed while the nut is screwed on.

The minor patent ("Deutsche Reichs-Gebrauchsmuster") for this type of insert barrel was filed as late as 5th March 1935. The corresponding patent was granted with DRGM number 1335123 on 25th April 1935.

The pre-production series of this variation was produced by refurbishing the old variation with thin locking nuts. In the process of this rework some of the muzzle markings were removed.

The later series production insert barrels were provided with the wider locking nut, a lock washer instead of a plain washer. This type corresponds with the PP manual dated 1936 and was produced largely unchanged until 1945. From about 1937 onwards there was a Walther banner etched onto the barrel.

Insert barrels were offered as sets in two variations in little boxes. It is not clear whether Walther or a subcontractor produced these sets.

- 4-mm-barrel: loading cartridge with a chamber for the small 4-mm-M20 cartridge (up to 1935 with eight, from 1935 with three loading cartridges), cleaning rod made of steel with a leather dish to protect the muzzle, ejector rod
- 4-mm-barrel: barrel with a chamber for the small 4-mm-M20 cartridge, which had to be removed from the chamber with a spoon-shaped carrier, cleaning rod made of steel with a leather dish to protect the muzzle, ejector rod

- 4-mm-barrel: like set a) or b), but without a box and or a cleaning rod.

To fit these insert barrels the slide has to be removed first. Then the small barrel is inserted from the rear into the original barrel to be held in the chamber. The insert is then fixed from the muzzle side with the knurled locking nut.

Normally the variation without the loading cartridges is a bit more precise when shooting it, as here the 4-mm-M20 cartridges are loaded directly into the barrel through the ejection port.

With the other variation the small 4-mm-M20 cartridge is inserted into a loading cartridge first. This one is then fed to the chamber with the aid of the original magazine. The front of the loading cartridge hereby immerses into the rear of the insert barrel.

After every shot the empty 4 mm shell has to be removed from the insert barrel resp. the loading cartridge with the help of the ejector rod.

During the development of these insert barrels Walther experimented with different twist angles, but stuck to the right hand twist. Finally they came back to the original twist angle.

Up to 1945 Walther insert barrels were mostly sold in black boxes – normally with a green sticker for the Walther PP ones, and with a red sticker for the PPK ones. There were other colors of the boxes reported though. The interior of the boxes varies, too. The insert barrels produced towards the end of World War Two were shipped in simple, green cardboard boxes with a stamped inscription.

Other manufacturers produced insert barrels for the Walther PP/PPK pistols, too. The sales were handled e.g. by GECO. These sets came in simple cardboard boxes without any Walther inscription or sticker. These non-Walther barrels lack the marking at the muzzle as well.

In 1950 Walther/Ulm made the effort of producing design drawings for these insert barrels (1st variation). But they never produced them. This was done by Lothar Walther in Königsbronn. So at least this business remained within the family.

These post-1945 insert barrels carry the marking at the muzzle (see above), but some of them have left hand twist. They are 10 mm resp. 15 mm longer than the pre-1945 ones in order to make them usable for other pistol types as well.

The post-1945 made loading cartridges are about 1 mm wider at the front. So they do not fit into the rear of pre-1945 insert barrels.

The cleaning rods are no longer made of steel (with a leather dish), but of brass.

From 1972 onwards these insert barrels were given an etched PTB-Number to meet the demands of the new German gun laws.

Both Walther/Ulm and Lothar Walther/Königsbronn sold these Lothar Walther made insert barrel sets – Lothar Walther/Königsbronn in little, green boxes whereas Carl Walther/Ulm preferred boxes with a wood grain pattern.

PP/PPK long barrel type:

PP 292 326P (5" barrel with both "Crown/N" and "Eagle/N" proof marks near muzzle)

PPK 999 678 (with a PP barrel)

PP long barrel and slide type:

PP 779 150, 779 153 (matte chrome-plated; checkered trigger, front and backstrap; barrel "Böhler Stahl Antinit", made on special order for Glaser/Zurich, SN range: 779 151 to 779 160 - 10 pistols?)

PP 966 773 (10 round magazine, longer grip and longer grip plates)

PP 961 410 (11 round magazine; longer grip plates)

PP target pistol type 1:

design similar to Walther model "Olympia" (type 1938), i.e. slide open to the top, however in caliber 9 mm kurz with

bottom magazine release, safety lever on left hand side of slide, smooth wooden grip plates
PP 835 201 (1935)

PP target pistol type 2:

longer grip and longer, wooden grip plates, adjustable rear sight, longer barrel, checkered backstrap, 10 round magazine
PP 991 026

PP experimental pistol type 1:

additional, long safety lever in front of left grip plate – cf. Walther Mod. 8
PP 942 559

PP experimental pistol type 2:

caliber 9 mm kurz, selector lever in rear of grip tang for single shot ("E" for "Einzelfeuer") or full automatic ("D" for "Dauerfeuer" = fully automatic), bottom magazine release, single row magazine, long safety lever in front of left grip plate – cf. Walther Mod. 8
PP 179 198P

PP experimental pistol type 3:

fully automatic as before, but double row magazine
PP 321 245P, 321 250P, 321 257P

PPK experimental pistols:

no safety and no space milled away for the safety lever on left hand side of slide, round firing pin (inserted into the slide from the rear and secured by a screw and a pin from the top of the slide), no signal pin, standard PPK frame, no proof marks, SN on frame overstamped with two triangles – experimental pistol for the development of the KPK?
PPK 319 312K, 319 342K

PPK sample pistols (?):

no slide legend, no Walther banner on grips, inscription "Mod. Z. B. ...." (on front gripstrap) – sample pistols for future licensed production?
PPK 853 969 ("Mod. Z. B.1054")
PPK 853 978 ("Mod. Z. B.1034")

Total production of Zella-Mehlis Walther PP/PPK pistols: about 830 900 pistols (540,000 PPs and 290,900 PPKs)

Walther PP Pre-1945 Copies

Walther PP Copy "Pistolet Warnant", France 1933

Dr. Dirk Ziesing, a gun expert working for the Bochum chamber of commerce, pointed out this interesting pistol "Pistolet Warnant – Breveté S.G.D.G. caliber 7.65 mm French Longue" (7.65 x 20 mm).

In 1933 George Warnant, Paris, handed in this pistol to the French army for assessment. At first sight it looks like an enlarged Walther PP only – or in other words: like a Walther "Militär-Pistole" – see there. The design characteristics according to the exploded view are almost identical with the ones of the 2nd variation Walther PP.

This pistol was designed for the caliber 7.65 longue, which was favored by the French army commission from 1927 onwards.

In 1935, however, a different pistol in this caliber, which was designed by the Swiss-born Karl Gabriel Petter, was introduced as a standard service pistol by the French army. So the attempt to place a Warnant product as a service pistol had failed. Whether Walther/ Zella-Mehlis after all knew about this whole procedure is questionable.

The dimensions of the "Pistolet Warnant" are: total length 8.0", barrel length 5.1", weight 31.7 oz., magazine capacity: 9 rounds 7.65 longue.

Walther PP Copy "Baltietz Pistol", Soviet Union 1942

This PP copy was born out of sheer necessity, as the occupation of Leningrad by German troops in the hard winter 1941/42 led to immense losses of material. The Soviets made a new experience, too: The Tokarev pistol, which was so far praised for its reliability and robustness, showed functional problems in extremely cold conditions – especially when condensation water froze. So the commander of the Baltic Soviet Fleet decreed the development of a more reliable pistol.

It was decided that a copy of a proven and popular blowback pistol, the Walther PP, should be produced. Already in those days this model was valued in the Soviet Union as a synonym for a reliable double action pistol. However this new pistol should be designed for the powerful cartridge 7.62 Tokarev – despite the unlocked blowback action. So one took the same route as Walther had done with the development of the first unlocked "Militär-Pistole" (cf. as well "Pistolet Warnant" – see above): all dimensions and weights were enlarged to compensate the strong blowback momentum of the Tokarev cartridge by the larger mass.

After some successful tests with "Baltietz Pistol No. 1" more tests were made with a slightly shortened and lighter "No. 2". The latter variation was then accepted for series production. There were, however, only 13 pistols made by hand, as there was a lack of raw materials and production machines. These pistols were issued to top party members and military officers in Leningrad – amongst others to Admiral Kuznezov, people's commissioner of the Soviet Navy.

Three specimens of this extremely rare Walther PP copy can be found today in the Central Armed Forces Museum in St. Petersburg (ex-Leningrad).

The dimensions of the "Baltietz pistol" are: total length 7.6", barrel length 4.7", weight 31.7 oz., magazine capacity: 8 rounds 7.62 Tokarev.

Walther PP and PPK Walther/Ulm Post-1945 Production Preliminary Remarks

After the 3rd US Army under General Patton had occupied Zella-Mehlis and the premises of Walther on 9th April 1945 the Soviets intervened politically. So the US troops were made to withdraw behind a line, which is roughly identical with the course of the later western border of the Soviet Occupation Zone.

So one often reads that end of June 1945 the US army took the Walther family to the Ulm area with nothing but the bare necessities and some hand luggage – without any company files (e.g. design drawings of the most important models).

This is, however, only partially true. Long before 1945 the Walther family owned a summer "residence" in Konstanz-Staad on Lake Constance. During World War Two this house was let to a former Walther foreman ("Waffenmeister"). As Fritz Walther, the Walther director, believed only superficially in the final victory of the Nazi regime, he made provisions for post-1945 times. He was clever enough to deposit copies of important files and design drawings in this summer house. In 1953 Fritz Walther had his house rebuilt already.

So the whereabouts of the remaining original company files after 1945 still need clarification. Obviously they were not in the possession of the ex-GDR government. Otherwise there would not have been a need in 1947 for twelve technicians to sit down in the August-Bebel factory and reconstruct the design drawings for the Walther PP. This was done for the later production of the pistol "Pistole 1001" – see there. If the Soviets had had those drawings in their possession they would presumably have handed them over to their "socialist brothers".

After Fritz and Erich Walther had produced mechanical calculators from October 1946 in Heidenheim/Brenz, they relocated the company to Ulm/Donau (on the River Danube) on 14th June 1950. There in the so-called "Donaubastion", an old barracks complex, they were allowed to produce air guns only – due to the regulations of the allied victors.

But in 1952 already the first post-1945 Walther PP was produced under license by "Manufacture de Machines du Haut-Rhin" (short: "Manurhin") in Bourtzwiler-Mulhouse (in German: Mülhausen). This place is situated on the upper Rhine River in Alsace – the traditionally bilingual part of East France speaking a German dialect as well as French.

In the Walther literature distinctions are often made between PP/PPK pistols made by Manurhin and those made by Walther/Ulm. Sometimes even the differing production quality is discussed in detail.

This differentiation is not appropriate, however, as up to the end of 1985 all PPs and PPKs were made by Manurhin – no matter what the slide legend says. Pistols with Manurhin slide legend were made under license and sold by Manurhin themselves or via Walther. Manurhin served those markets, which were either traditionally linked with France or which could not (yet) be served by German companies due to legal or other restrictions – e.g. West Berlin.

For example in 1957 Manurhin PPKs were delivered to the British secret service MI6 (from about SN 126 000 onwards). Hence the PPK was the service pistol of James Bond, too. From 8th January 1966 Manurhin shipped 3 344 PPKs to the West-Berlin police, which was not allowed to procure "German" guns due to an Allied Control Council Act. As late as in the 1980s these pistols were taken out of service again.

In 1984 Manurhin was taken over by the French Matra group. Hereafter there was a legal dispute between Matra and Walther about the naming and license rights. This finally led to an end of this business relationship. Manurhin today produces mainly machinery for the production of ammunition.

Manurhin started the production of Walther PPs (from SN 10 001) in 1952 and from 1954 they produced Walther PPKs (from SN 100 001). Those PP/PPK pistols, which were made by Manurhin to be sold under license, have a Manurhin slide legend. PP/PPK slides, which were later given a Walther/Ulm inscription were shipped to Ulm in-the-white and unhardened. There they were slightly refurbished, roll-stamped with the inscription, hardened, polished and finally salt blued. Frames were delivered to Ulm with the blued finish already. The final assembly of Walther/Ulm PP/PPK pistols was done in Ulm. There they were supplied with grips and magazines, proofed and then shipped to the customers.

The first PPs/PPKs made by Manurhin, but given the Walther/Ulm slide legend, were:

PP	10 003 (7.65 mm Browning)
PP	12 025 (9 mm kurz)
PP	10 002 (.22 l.r.)
PPK	100 001 (7.65 mm Browning)
PPK	100 305 (9 mm kurz)
PPK	100 055 (.22 l.r.)

Only after the separation from Manurhin and after the license contract had run out in 1986 did Walther/Ulm start its own production of Walther PP/PPK pistols.

But **Walther/Ulm** had made the first test productions before already. These pistols had the prefix "V" in their serial numbers ("V" for "Versuch", in English "experiment"):

- On 30th October 1980: one Walther PPK in caliber 7.65 mm Browning (V1197),
- On 13th November 1982: one PP in caliber 9 mm kurz (V1321) and
- On 4th December 1984: three Walther PPKs in caliber 9 mm kurz (V1365 – V1367).

Manurhin as well as Walther/Ulm produced PPs/PPKs in the calibers .22 l.r., 7.65 mm Browning (.32 ACP) and 9 mm kurz (.380 ACP) only. Pistols in caliber 6.35 mm Browning (.25 ACP) were no longer made. There were no Dural frames made for PPs in all the three calibers and for PPKs in caliber 9 mm kurz.

All PPs/PPKs produced by Walther/Ulm in 1986 were given the additional inscription "100 Jahre Walther" on the right hand side of the slide.

It is hardly known that Manurhin produced frames and slides for another major German manufacturer, too. These parts were produced by Manurhin and shipped to Germany in-the-white. There these major parts were finished and assembled using minor parts from other subcontractors.

Serial Number Ranges Walther PP/PPK (post-1945 production)

All PPs and PPKs either with Manurhin or Walther slide legend were numbered in the same serial number ranges.

On the frames the serial numbers were applied vertically with the base to the grip plates. The very first post-1945 PPs and PPKs, however, were stamped with the serial numbers "upside- down" – cf. the early PPs and PPKs of the Zella-Mehlis production (see there). From 1965 onwards the serial number was additionally stamped on the right hand side of the slide. In both cases roll-stamps were used. At the customer's (e.g. some authorities) request the serial number (or at least its last three digits) was optionally applied to the slide before 1965, too.

From the beginning of 1986, when Walther started its own production, new serial number ranges were used: PPs from SN 700 001 and PPKs from SN 800 001.

Numbers with six or seven digits (PPs) or letters (PPKs) at the bottom of the frame (below the grip plates) are internal batch codes for the licensed production – independent of the type of pistol or the caliber.

With post-1945 PP/PPK pistols in the calibers 9 mm kurz and .22 l.r. suffixes were added to the serial numbers: "A" for 9 mm kurz (no longer true for Walther production after 1986!) and "LR" for .22 l.r.

The suffixes "R" or "D" were presumably used when accidentally there were double assignments of serial number ranges.

The suffix "L" on the underside of the slide (!) designates pistols for left-handed shooters with a modified safety (see picture 30).

The target versions of the Walther PP were produced with short or long barrels. The ones with short barrels were given the suffix "C" (for French "court" = short) and the ones with long barrels were given "L" (for French "longue" = long).

Faulty slide designations:

Some PPKs in caliber 9 mm kurz in the SN range around 160 400 were erroneously marked "Mod. PP" on the left hand side of the slide.

Grip Plates Walther PP/PPK (post-1945 production)

As a rule **post-1945 PPK** grips are two-piece ones. Early Manurhin PPKs, however, still have honey-colored to dark-brown one-piece wrap-around grips without a metal reinforcement in the lower backstrap (cf. pre-1945 PPKs) and the Manurhin emblem on both sides.

Later PPKs made by Manurhin or Walther/Ulm have two-piece grips made of "Cellidor" ("Cellidor" is an injection-molded plastic based on cellulose acetate) with a checkered outside. The company emblem is either the Manurhin logo (with a mill wheel) or the Walther banner. The color up to 1980 was a mottled brown, from then onwards black. White grip plates were optional at an extra charge.

Post-1945 PPs have two-piece grip plates, which resemble the ones of the 4th variation Zella-Mehlis PPs – see there. The Walther banner is smaller though.

These grip plates, too, were made of "Cellidor" using injection molding technique. The company emblem is either the Manurhin logo (with a mill wheel) or the Walther banner. The color is black (very rarely a mottled dark-brown). White grip plates were optional at an extra charge.

White grip plates or the ones made of special material (ivory, carved wood etc.) were mostly used with engraved PPs/PPKs at an extra charge.

Some early Manurhin PPs (mostly target pistols) came with a light thumb rest on the left grip plate – often combined with the "boxed" magazine extension.

Magazines Walther PP/PPK (post-1945 production)

Magazines for PPs/PPKs in caliber .22 l.r. have no inscription on the body, but a button in the viewing slot on the right hand side of the body and only one hole on the left hand side of the body. This button works as a loading device. Magazines for Manurhin PPs/PPKs in caliber .22 l.r. have the inscription "France" on the right hand side of the body. The follower is made of pressed sheet metal and activated by a zigzag spring.

Magazines for PPs in caliber 7.65 mm Browning have either the Walther banner or the Manurhin emblem on the left hand side of the body plus the model and caliber designation – "PP 7,65 mm" (Note the spelling of "mm" without a slash – cf. Zella-Mehlis PPs).

From about 1968 (introduction of the model PPK/S) PP magazines show no model designation, but the caliber designation "Cal. 7,65 mm". Some post-1945 PP magazines have no inscription at all.

What they all have in common is: seven viewing holes on both sides of the body.

From about 1968 onwards the upper front of the magazine body was no longer straight, but with a curved cutout (cf. magazines for 9 mm PPs/PPKs). This modification was made to improve the feeding of the 7.65 mm cartridges and to simplify production, as the magazines for 9 mm PPs/PPKs had this cutout anyway.

Magazines for PPKs in caliber 7.65 mm Browning, too, have either the Walther banner or the Manurhin emblem on the left hand side of the body plus the model and caliber designation – "PPK 7,65 mm" (Note the spelling of "mm" without a slash – cf. Zella-Mehlis PPKs).

These PPK magazines have six viewing holes on both sides of the body.

From about 1968 onwards the upper front of the magazine body of PPK magazines in this caliber was no longer straight, but with a curved cutout (cf. magazines for 7.65 mm PPs and 9 mm PPs/PPKs).

Later PPKs and PPK/S pistols (especially for the US market) have the straight upper front again.

Magazines for Manurhin PPs/PPKs in caliber 7.65 mm Browning have the additional inscription "Made in France" on the right hand side of the body.

Magazines for PPs/PPKs in caliber 9 mm kurz have a longitudinal rib on the left hand side of the body in order not to mix them up with magazines for pistols in caliber 7.65 mm Browning. Other than that the inscription on the left hand side of the body is as described above – although with the caliber designation "Cal. 9 mm kurz" or "Cal. .380 ACP". The "maximum version" of inscriptions is: Walther banner on both sides, left hand side with "Cal. .380 ACP", right hand side with "Cal. 9 mm kurz" plus all the viewing holes numbered.

Some PP/PPK magazines in this caliber have no inscription at all, however.

Magazines for PPs/PPKs in caliber 9 mm kurz have the crescent-shaped cutout at the front of the body for the upper magazine release. The front of the magazine body shows a curved cutout at the top for an improved feeding of the 9 mm cartridges – cf. 7.65 mm magazines.

Magazines for PPs in 9 mm kurz have six viewing holes, whereas the ones for PPKs have only five viewing holes on the right hand side of the magazine body. In both cases there is only one hole on the left hand side.; remark: This single hole on the left hand side of the body of magazines is not a viewing hole. It was used during production as a reference hole for positioning the magazine body precisely in the machine.

Magazine Extensions Walther PP/PPK (post-1945 production)

The magazine extensions for post-1945 PPs/PPKs resemble the ones made for the 5th variation Zella-Mehlis PPs/PPKs: recess at the bottom and on both sides, the front is drawn down with a thick rim. The contours of the recesses at the sides and at the bottom are more angular though.

The color of the PP magazine extensions is black. The one of the PPK extensions is rarely brown, but mostly black, too. White extensions are rare and mostly used together with white grips. All colors were available as options, too.

Longer (so-called "boxed") extensions were offered as an option for the Manurhin PPs, too – cf. Zella-Mehlis PP magazine extensions.

Walther PPs/PPKs Zella-Mehlis/Ulm "Hybrids"

One special variation of Walther PPs/PPKs has often puzzled some experts: Those pistols have serial numbers and all the characteristics of Zella-Mehlis PPs/PPKs. The slide legend, however, contains neither the place name "Zella-Mehlis" nor "Ulm/Do.". These pistols are mostly in a very good condition.

The first thought normally is that these pistols were assembled from Zella-Mehlis parts mixed with some from Ulm production. Even then they should carry one of the two place names in the slide inscription. The solution to this puzzle is that in the 1950s to the 1980s Walther/Ulm refurbished some older PPs/PPKs at the customer's request. These pistols were sanded and polished, if necessary completed with new parts. Before blueing they were given a new slide legend with a roll-stamp, which lacked the place name "Ulm/Do.". Or new slides were used - again without any place name in the legend.

Walther PP Manurhin and Walther/Ulm

Picture 24: *Walther PPK ZM/Ulm, SN 375 408K, caliber 7.65 mm Browning (.32 ACP), made in 1942 by Walther Zella-Mehlis and refurbished by Walther/Ulm in 1969*

Production:	1952 – 1999 ("Last edition")
Calibers:	7.65 mm Browning = .32 ACP (8 rounds) 9 mm kurz (9x17) = .380 ACP (7 rounds) . 22 l.r. (9 rounds)
Total Length:	6.7 in.
Barrel Length:	3.6 in.
Weight:	22.9 oz. (lightweight) or 23.7 (steel)
Rifling:	6 grooves, right-twisted
Operation:	unlocked blowback action with firing pin and exposed hammer, double action/ single action trigger mechanism

Characteristics: as 5th variation PP Zella-Mehlis, but

- Slide legend left hand side: "MANUFACTURE DE MACHINES DU HAUT-RHIN / MADE IN FRANCE / LIC. EXCL WALTHER MOD. PP Cal. 7,65 mm" resp. "9 mm kurz" or ".22 l.r." or: "Carl Walther Waffenfabrik Ulm/Do. / Mod. PP Cal. 7,65 mm" resp. "9 mm kurz" or ".22 l.r." with Walther banner on left hand side of inscription; Remark: The spelling of "mm" on post-1945 PPs and PPKs is without a slash.
- PPs/PPKs made for export up to 1968 had the additional inscription "Made in W. Germany" vertically on the frame; all PPs from 1968 carried "Made in W. Germany" horizontally on the slide.

Picture 25: *Walther PP, Manurhin slide legend, SN 34 403, caliber 7.65 mm Browning (.32 ACP), produced by Manurhin in 1955*

- Sam Cummings from Interarmco, later Interarms, was the biggest importer of Walther pistols in the US. His predecessor had been Thalson Co., San Francisco. From the 1950s until 1961 Manurhin produced PP/PPK pistols for Interarms marked with the Walther banner, "Made in France" and "Mark II", as Sam Cummings wanted the Walther banner due to its drawing power. From 1961 Interarms received PP/PPK pistols, which were marked with the Walther banner anyway. So the additional designation "Mark II" disappeared. Allegedly Manurhin produced PPs for Interarms again from 1981 to about 1985, when the US production suffered from starting problems. These pistols were marked "Made in France" again. After the takeover of Manurhin in 1984 by Matra the slide legend said "Matra Manurhin International Ft. Lauderdale FL".
- Slide legend right hand side: mostly none – except for pistols exported to the US. These are marked "Interarms Alexandria Virginia" with the company logo. From about 1965 the serial number was applied on the right hand side of the slide, too.
- Proof marks: either French proof marks from Paris proof house ("ship with three sails") resp. St. Etienne ("St. Etienne" with a crown) or German proof marks from Ulm proof house ("Federal Eagle over N" with an "antler", from 1961 with the end digits of the year, from 1977 with the year code – see appendix)
- The French proof marks put on PPs made in the 1950s often were "Crown with a Shield over a Mill Wheel". This was the stamp for the Mulhouse (German: "Mülhausen") proof house, a branch of the St. Etienne proof house.

Remarks:

In the US literature this emblem is often interpreted as a "cog wheel" as associated with machinery. In fact this is a mill wheel, as the German place name "Mülhausen" is derived from "Mühle", meaning mill, i.e. a facility to grind grain.

The proof marks described above can sometimes be found in a mixed constellation on one pistol.

- Frames are of the "closed", reinforced type (like e.g. Zella-Mehlis PPs in caliber 9 mm kurz – see there and illustration 2). Post-1945 PP frames were no longer made of "Dural".
- Frames have the milled step at the trigger guard pin, but the long grip tang (like the 5th variation Zella-Mehlis PPs – see there).
- Sear: asymmetrical – see illustration 2
- Sights: guilloché slide rib with milled front sight and adjustable rear sight (inserted in a dovetail)
- Magazine release: push button on upper left hand side of frame – with PPs in all calibers
- Width of safety lever: "slim" lever (5 grooves)
- Swiveling range of safety lever: 60°
- Firing pin: one-piece firing pin without a housing, underside of slide is milled through to get access to the firing pin and the signal pin; with PPs in caliber .22 l.r. this underside is not milled through, however, as there cannot be a signal pin because of the rimfire cartridge
- Signal pin (i.e. loaded chamber indicator): no longer optional, but standard (except for PPs in caliber .22 l.r.); now made of round material with a cutout in the middle for the removal tool and with a modified spring
- Ejector: pivoting 5° upwards; ejectors for PPs in caliber .380 no longer have an additional "spur" at the bottom
- Metal finish: high-gloss polished finish as standard finish; chrome-, gold-, silver- or nickel-plated finish or engravings (with blued, gold-, silver- or nickel-plated finish) optional at an extra charge; in 1988 Walther made test runs with a matte blued finish (glass bead blasting before blueing)
- Lanyard ring: optional item (when installed it replaces the main spring plunger) with a swiveling ring, base of ring more angular than the one for Zella-Mehlis PPs
- Grip plates: two-piece grips made of black "Cellidor" plastic (see above); outside checkered; Manurhin logo or Walther banner; white or carved wood grip plates optional at an extra charge

Total production of Walther PP pistols Manurhin and Walther/Ulm: about 400,000 pistols (1952 - 1999)

Walther PPK Manurhin and Walther/Ulm

Production:	1954 - 2003
Calibers:	7.65 mm Browning = .32 ACP (7 rounds) 9 mm kurz (9x17) = .380 ACP (6 rounds) . 22 l.r. (8 rounds)
Total Length:	6.1 in.
Barrel Length:	3.3 in.
Weight:	16.9 oz. (lightweight) or 19.2 (steel)
Rifling:	6 grooves, right-twisted
Operation:	unlocked blowback action with firing pin and exposed hammer, double action/ single action trigger mechanism

Characteristics: as 5th variation PPK Zella-Mehlis (see there) resp. post-1945 PP, but

- Slide legend left hand side: "MANUFACTURE DE MACHINES DU HAUT-RHIN / MADE IN FRANCE / LIC. EXCL WALTHER MOD. PPK Cal. 7,65 mm" resp. "9 mm kurz" or ".22 l.r." or: "Carl Walther Waffenfabrik Ulm/Do. / Mod. PPK Cal. 7,65 mm" resp. "9 mm kurz" or ".22 l.r." with Walther banner on left hand side of inscription;
- Slides: Zella-Mehlis PPKs have a stepped contour at the bottom of the slide (below the muzzle); post-1945 PPK slides have a curved contour there (cf. PPK prototype and very early Zella-Mehlis PPKs, too – see there).
- Lanyard loop: punched and bent steel item at the heel of the grip, serves as the main spring counter bearing
- Grip plates: two-piece grips (except for early Manurhin PPKs – see above) made of brown resp. later black "Cellidor" plastic ("Cellidor" is an injection-molded plastic based on cellulose acetate); outside checkered; Manurhin logo or Walther banner; white or carved wood grip plates optional at an extra charge
- PPKs in .22 l.r. and 7.65 mm Browning were also made with lightweight frames (s. chapter "PPK-L").

Total production of Walther PPK pistols Manurhin and Walther/Ulm: about 633,000 pistols (1954 - 2003)

Picture 26: *Walther PPK, Manurhin slide legend, SN 103 887, caliber 7.65 mm Browning (.32 ACP), produced by Manurhin in 1955 for the Austrian custom authorities*

Walther PP and PPK Manurhin and Walther/Ulm Development and Technical Changes

PP/PPK from about 1968:

- Sights modified: box-shaped rear sight inserted into a dovetail in slide rib; mostly thicker front sight with contrasting paint

PP/PPK from about 1971/72:

- Firing pin modified: rear end no longer with a lateral lug, but with a ball-shaped bulge; safety shaft and plunger are modified accordingly

PP from SN 422 081 (caliber 7.65mm Browning)
from SN 46 261A (caliber 9 mm kurz)
PPK from SN 271 511 (caliber 7.65mm Browning)
from SN 163 027A (caliber 9 mm kurz)

PP/PPK from about 1973/74:

- Hammer block modified: chamfer at the bottom was omitted

PP/PPK from about 1986:

- Frame modified: lateral chamfer at the bottom of the slide (both sides, above the trigger guard) was omitted
- PP from about SN 710 000
- PPK from about SN 810 000

Walther PPK-L Manurhin and Walther/Ulm

Picture 27: *Walther PPK-L, Walther/Ulm slide legend, SN 521 745, caliber 7.65 mm Browning (.32 ACP), produced by Manurhin in 1967*

Production:	March 1954 - January 1982
Calibers:	7.65 mm Browning = .32 ACP (7 rounds) . 22 l.r. (8 rounds)
Total Length:	6.1 in.
Barrel Length:	3.3 in.
Weight:	22.9 oz.
Rifling:	6 grooves, right-twisted
Operation:	unlocked blowback action with firing pin and exposed hammer, double action/ single action trigger mechanism

Characteristics:

- Frames are of the "closed", reinforced type (like e.g. Zella-Mehlis PPKs in caliber 9 mm kurz – see there and illustration 2)
- Frames have no milled step at the trigger guard pin and the long rear grip tang (like the 6th variation Zella-Mehlis PPKs with a steel frame – see there). Remark: Late Zella-Mehlis PPKs with Dural frames have a short rear grip tang, but no milled step at the trigger guard base.
- Sear: asymmetrical – see illustration 2
- Barrel pressed into the frame from the rear, secured by a collar and a pin – cf. late Zella-Mehlis Dural PPKs
- Slide legend: early PPK-L pistols from 1954 with Walther slide legend: "Mod. PPK Cal. …" (without suffix "L"); later pistols: "Mod. PPK-L Cal. …" (with suffix "L"); PPK-L pistols with Manurhin slide legend have no suffix "L" on the slide, only the box is marked "Dural"
- Some PPK-L pistols ("P21") for the German army ("Bundeswehr") have the slide legend "Modell PPK Kaliber 7,65

mm L" (Note the spelling of the caliber designation and the position of the suffix – see "P21" as well)

- About 100 PPK-L pistols were stamped on the frame with the suffix "L" to the serial number.
- In the SN range 126 679 – 127 078 [sic!] PPK-L pistols were numbered in the PPK serial range (October 1968).
- PPK-L pistols were not produced in caliber 9 mm kurz.
- PPK-L pistols have their own serial number range:

SN range PPK-L: 500 001 - 533 640

Remarks:

All PPK-L pistols were produced by Manurhin in Mulhouse – including the ones with Walther/Ulm slide legend.

The very first PPK-L with SN 500 001 (caliber 7.65 mm Browning) went to the Canadian importer of Walther pistols. The one with SN 500 001LR (caliber .22 l.r.) went to the US importer (Interarms).

Total production of Walther PPK-L pistols Manurhin and Walther/Ulm:
33,640 pistols

Walther PPK Licensed US Production

PPK Production Interarms / Carl Walther USA / Smith & Wesson

Next to the PPs/PPKs, which were produced by Manurhin under license, there was a licensed US American production of the Walther PPK, too. This was especially important, as after 1968 the import of Walther PPKs was forbidden due to the new Gun Control Act and the fact that the high US dollar exchange rate made exports to the US difficult.

Samuel Cummings from Interarms in Alexandria/Virginia, the US importer of Walther pistols, had the Walther PPK produced for his company. From 1986 the US PPKs came in caliber .380 ACP (9 mm kurz) made of stainless steel (SN with prefix "A"). From 1986 they were made of blued steel in caliber .380 ACP (SN with prefix "K") and from 1998 they were made of stainless steel in caliber .32 ACP (7.65 mm Browning), too (SN with prefix "B"). The slide legend carries the Interarms company address on the right hand side of the slide.

Interarms did not produce these pistols itself, but had them manufactured first by Ranger Manufacturing Co., then by Black Creek Inc. – both were parts of the company consortium Mid South Industries in Gadsden/ Alabama. In 1999 the license contract with Interarms expired and Carl Walther USA LLC became the new producer together with Smith & Wesson.

On 1st January 2013 Walther Arms Inc., Fort Smith/Arkansas (the newly founded subsidiary of Walther/Ulm), took over the sales.

The new PPK/S in caliber .22 l.r. is produced by Walther/Arnsberg.

Characteristics of the US PPK:

as the PPK produced by Walther/Ulm, but

- Slide is a bit more "full-bodied" near the muzzle.
- Frame not forged but investment casted
- No lanyard loop

- Rear grip tang longer and more pointed – often named "beavertail" in the US literature
- Slide legend with the additional inscription "Under License of Carl Walther Ulm/Do." on pistols made for Interarms, resp. " … License" …" on those made by S&W
- From 1986: made of stainless steel or blued steel, in caliber .380 ACP
- From 1998: additionally made of stainless steel in caliber .32 ACP, too

Walther PP, PPK, PPK-L Manurhin and Walther/Ulm Contracts and Special Models

Picture 28: *Walther PPK ("P21"), Walther/Ulm slide legend, SN 132 103, caliber 7.65 mm Browning (.32 ACP), produced by Manurhin in 1958 for the German "Bundeswehr", marked "Bw", high-gloss polished blued finish*

a) Walther PPKs/PPK-Ls ("P21") and PPs ("P22") for the post-1945 German army ("Bundeswehr"):

By approval of a decree dated 22nd November 1956 the Walther PPK was introduced as a service pistol ("Pistole 7,65 mm x 17 P21" or short "P21") – first for the German air force only and from 1966 for the army and navy, too. Hence the military classification as "Luftwaffengerät" (air force equipment) first. To start with the P21 was subject to the air force material management. In November 1977 the material management of the P21 was handed over to the army, however.

Already on 18th April 1956 Walther shipped 243 pistols P21 (PPK-L type) as

PPKs. Due to their slide legends these pistols were regarded as "French" weapons.

These West-Berlin service pistols were marked on the frame with the so-called "Berlin Rosette" – actually a stylized police star. They were taken out of service from 1980 onwards.

The investigation unit of the German railway police procured Walther PPKs from 1979 onwards, whereas the uniformed railway police preferred the Walther PP. These pistols were marked with "DB" (for "Deutsche Bundesbahn" – not to be confused with weapons of the Daimler-Benz factory security, which were marked similarly).

Uniformed railway police officers carried Walther PP pistols.

The "Bundesfinanzministerium" (Federal Ministry of Finances) acquired Walther PPs and PPKs for the German customs authorities. Those were marked "BUND".

The "Bundesministerium des Innern" (Ministry of the Interior) procured Walther PPs and PPKs as well for their subordinate authorities, e.g. the "Bundesgrenzschutz" (Federal Border Guard) – marked with "BMI".

The "Deutsche Bundespost" (Federal Postal Service) acquired Walther PPKs, too, but in contrast to the pre-1945 ones these pistols were not marked – see PPK "D.R.P."

From 1952 onwards the new "Bundesregierung" (Federal Government) equipped its ministers, secretaries of state and high-ranking officials with Walther or FN pistols for their self-defense. So the Federal Foreign Office bought 21 PPKs, which were not marked especially.

SN range:
101 338 - 163 364 (SNs are not consecutive!)

The "Bundesministerium für Ernährung, Landwirtschaft und Forsten" (Ministry of Agriculture and Forestry") bought Walther PPKs from 1955 onwards, which were taken out of service as late as 2011. These pistols were not marked especially.

SN range:
105 617 - 133 269
(SNs are not consecutive!)

c) Walther PPKs for the Swiss Air Force ("Fliegerpistole"):

In 1965 the Swiss Air Force (then still called "Flugwaffe" or "Fliegertruppen") procured 415 PPK-L pistols (SN range 513 796 – 515 938) plus four PPKs with steel frames (all in caliber 7.65 mm Browning).

In a second delivery the Swiss received further 315 PPKs (with steel frames) in the SN range 210 000 – about 280 000.

These pistol were given the Swiss military designation "Selbstladepistole, Ordonnanzmodell 1965" (self-loading pistol, ordnance model 1965). They were issued to the "Fliegerstaffeln" (flying squadrons) as "Korpsmaterial", i.e. equipment, which was stored by the military corps and handed out to the pilots only directly before a flight duty.

The PPK-Ls were mainly used by air pilots, whereas the steel PPKs were issued to ground and military judicial staff.

The military supply number for the PPKs was 1005-670-5170 and for the PPK-Ls it was 1005-670-5171.

From 1992 onwards these pistols were taken out of service by the Swiss "Kriegsmaterialverwaltung" (war material administration).

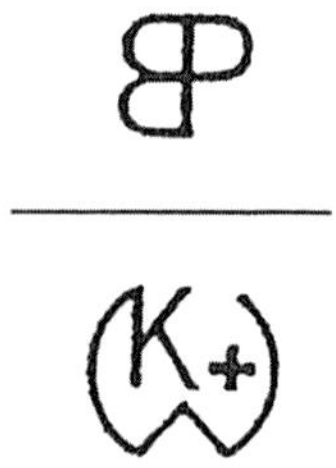

Illustration 3: *Swiss proof mark and acceptance stamp*

Markings:

Next to the commercial German proof marks these "Fliegerpistolen" have on the left hand side of the slide the Swiss proof mark and mostly on the rear face of the chamber a reversed "B" with a "P" for the so-called "Tormentierbeschuss", i.e. test firing with a pressure stress cartridge). On the right hand side of the frame there is the acceptance mark of the Swiss "Eidgenössische Konstruktionswerke" in Thun/Switzerland: a "K+" embedded into a "W" – see illustration 3.

d) Walther PPs and PPKs for Austrian Authorities:

The Austrian "Sicherheitsbehörden", Police and "Gendarmerie" (constabulary or rural police), procured Walther PPs and PPKs and marked these on the left hand side of the frame with the Austrian Federal Eagle as an acceptance and a property mark. The magazines were stamped with the serial number and an additional "I" or "II".

The so-called "Sicherheitswache" (short "SW"; security guard) was up to 2005 a uniformed police force in charge of maintaining public safety and order. Meanwhile the SW is part of the Austrian federal police. This police force, too, purchased Walther pistols (mostly PPs) and marked them with "SW" (and inventory number) on the left hand side of the frame – plus the Austrian Federal Eagle. Only officers carried a PPK, then marked "SWO"

The plain clothes detctives of the "Bundesgendarmerie" were in charge of rural areas and subdivided in different "Landesgendarmeriekommandos" ("LGK"). They carried Walther PPKs marked "LGK" plus the abbreviation for the federal state and an inventory number.

The "Kriminalbeamtenkorps", which existed next to the "Sicherheitswache", was in charge of urban areas and procured PPKs - marked "POL".

The Austrian "Justizwache" (short "JW"; judicial authorities guard) is in charge of guarding courts of justice and prisons. This police force, too, purchased PPs/PPKs and marked them with "JW" (and inventory number) on the grip – plus the Austrian Federal Eagle.

The Austrian customs authorities, too, purchased Walther pistols (mostly PPKs) and marked them with "Z" (and inventory number) on the left hand side of the frame – plus the Austrian Federal Eagle (cf. picture 26).

The Austrian army ("Bundesheer") procured Walther PPKs to arm their pilots, officials of the Ministry of Defense and intelligence service – marked with the acceptance stamp of the "Amt für Wehrtechnik" (Eagle over BHI).

e) Walther PPs and PPKs for the Swedish Police:

As they had done before 1945 already, the Swedish police procured Walther pistols (mostly PPs) in post-1945 years, too. They marked these on the right hand side of the slide with "Tillhör Polisen" (Swedish for "property of the police"). Most of the Swedish PPs were supplied with the swiveling factory lanyard ring at the heel of the grip.

f) Walther PPs and PPKs for the Danish National Police ("Rigspolitiet") and other Forces:

From 1966 the Danish National Police ("Rigspolitiet", short: "Rplt.") procured post-1945 Walther PP/PPK pistols to be used together with the Zella-Mehlis made ones, which were already in their arsenals. From 1945 up to 1966 the Danish uniformed police was still unarmed.

In addition to these stocks there were the ones taken over after 1945 e.g. from old German "Wehrmacht" arsenals. Some of these Walther PP/PPK pistols were in caliber 9 mm kurz and had the bottom magazine release.

These pistols were converted to caliber 7.65 mm Browning, but kept their bottom magazine release.

The Danish police not only purchased complete pistols, but also parts like slides and frames. With these they replaced broken parts from Zella-Mehlis pistols. Hence there are some "hybrid" PPs and PPKs as well.

As a rule the "Rigspolitiet" marked their pistols with "Rplt." (plus inventory number) on the left hand side of the frame. In post-1945 days this was obviously not done consistently any more. So there are some Danish National Police PPs and PPKs known without this marking.

Most of the magazines were subsequently marked by police armorers with the serial number of the related pistol – at least with the last three digits – at the side of magazine extension.

After forty and sometimes more years of service and presumably due to the use of a more powerful ammunition, some "Rigspolitiet" PPs and PPKs showed cracks in the slide. After a team from Walther/Ulm had inspected these pistols, there was an agreement in 1988:

Walther/Ulm provided 1500 reinforced replacement slides. Basically these slides were not reinforced, however, by adding any reinforcing material. There were simply some steps in production omitted, which weakened the slides. The signal pin as a loaded chamber indicator was eliminated. So the underside of the slide does not have to be milled through to give access to this pin. This underside remains massive and works as a cover for the firing pin (cf. PPs and PPKs in caliber .22 l.r.). A round firing pin (standard in series production from 1971/72) was used. So a cylindrical bore from the rear of the slide sufficed.

These replacement slides have the new box-shaped rear sights and the new, thicker and longer front sights. The slides were proofed by the Ulm proof house in 1988 and have the appropriate proof marks: "Federal Eagle over N", "antler" (for Ulm proof house) and the year code "JJ" for 1988.

Not only the Danish National Police, but also the Danish Border Guard procured Walther PPs and PPKs and marked them with "Gg".

The Danish Navy purchased Walther PPs and PPKs, too, and marked them with "SA" (over a crown).

g) Walther PPs and PPKs for the Israeli Army:

In the 1970s the Israelis equipped their armed marshals on aircrafts ("sky marshals") with Berettas Mod. 70 resp. 71 or Walther PPKs (all in caliber .22 l.r.) to protect their airline "El Al". They used a special type of ammunition with hollow-point bullets to make sure that in case of a shooting the outer shell of the airplane could not be pierced. This would have had disastrous effects for the pressure inside the cabin.

As these plain clothes sky marshals as a rule were members of the Israeli army – a unit called "Sajeret Matkal" – these pistols were marked with the Star of David on the frame (with the Hebrew letter for "tsadi" = Israel Defense Forces and the sign for "nun" = Israel proof mark) – see illustration 19 on page 177.

h) Walther PPs and PPKs for the British Army and Police:

The activities of the IRA (Irish Revolutionary Army) in Northern Ireland from 1969 onwards had the effect that the members of the British Army and police had to carry concealed personal defense weapons. This was especially true for the members of the "Ulster Defence Regiment" ("UDR") and their family members.

They chose the Walther PP and PPK (later replaced by the Walther P5 compact – see there), although the standard service pistol of the British Army was the FN HP9 (British army designation: "L9 A1").

Picture 29: *Walther PP, Walther/Ulm slide legend, SN 37 090LR, caliber .22 l.r., produced by Manurhin in 1973 for the British Army ("Personal Defence Weapon"), with "Suncorite" finish and lanyard ring*

In the mid-1970s the "Royal Army Ordnance Corps" bought about 3000 Walther PPs as "Personal Protection Weapons" ("PPW"). They received the following military designation: first "XL47E1", later "L47A1" for the PPs in caliber 7.65 mm Browning and "L66A1" for the ones in caliber .22 l.r. The letter "L" in these designations stands for "landforces".

PPK pistols were purchased in very limited numbers only. So they received the same military designation as the Walther PP.

The jet pilots of the Royal Air Force were also issued Walther PPs as emergency weapons.

The members of the URD and their family members got so used to carrying a defense weapon even in their spare time, that they referred to these pistols as "disco guns".

Walther/Ulm delivered these PPs with the standard high-gloss blued finish. Later these pistols were adapted to the British REME standards by British armorers: slides and frames were sandblasted and then sprayed with matte-black "Suncorite" lacquer. Some PPWs escaped this procedure, however, and still display their original high-gloss finish.

As late as January 1975 all PPWs in caliber 7.65 mm Browning were modified by fitting the new, round firing pin and the appropriate safety and plunger – although this modification was standard in the Walther series production from 1971/72 already.

From January 1980 most of the PPWs were retrofitted with swiveling lanyard rings, not always with the factory ones.

From January 1980 Walther PPs in caliber .22 l.r. were fitted with a new, reinforced firing pin and marked "P" (for "pin") on the rear of the slide.

The procuring of these PPWs was done in such a hurry, that these pistols did not pass the usual British military acceptance procedure. So they were not marked with the "broad arrow" (a broad, three-pronged arrow) – the usual British

Picture 30: *Walther PPK, Walther/Ulm slide legend, SN 245 179, caliber 7.65 mm Browning (.32 ACP), produced by Manurhin in 1970, modified for left-handed shooters with the safety on the right hand side of the slide*

Army property stamp. In cooperative assistance these PPWs were checked and proofed by the German "Güteprüfdienst des Bundesverteidigungsministeriums" (quality inspectorate of the Ministry of Defense). They were stamped on the left hand side of the frame with two of their acceptance stamps, which are often referred to as "Eagle with Drooping Wings".

From 1989 these PPWs were gradually taken out of service and replaced by Walther P5 compacts.

i) Walther PPs and PPKs for the Swiss Police:

Already before 1945 some Swiss canton and municipal police forces had purchased Zella-Mehlis made Walther PPs and PPKs – strangely enough some of them in caliber 6.35 mm Browning (.25 ACP), which is actually atypical for police service pistols.

Furthermore from 1969 onwards the Swiss airline "Swissair" used plain clothes police officers as "sky marshals" (in Swiss police terms unofficially called "Tigers"). Their standard service pistol was the Walther PPK in caliber .22 l.r. – the more so as the standard service pistol of most Swiss police forces was the Walther PPK in caliber 7.65 mm Browning anyway.

The first Walther PPKs for these "Tigers" were simply purchased from gun shops in and around Zurich/Switzerland, whereby it was not important whether these pistols had a Dural or a steel frame. Later the Swiss police purchased PPKs directly from Walther/Ulm.

Initially there were only 50 PPK pistols in their arsenals. Gradually this number was increased to about 150 PPKs.

From 1969 to 1997 these Swiss "sky marshals" were armed with Walther PPKs resp. PPK-Ls in caliber .22 l.r. Only some of these (especially the PPK-Ls) had consecutive serial numbers. The ammunition used were Remington cartridges with hollow-point bullets.

In 1997 the Walther PPK resp. PPK-L pistols for the "Swissair Tigers" were replaced by SIG-Sauer Mod. P232 pistols in caliber 9 mm kurz (.380 ACP).

With a few exceptions the Walther pistols were destroyed after they were taken out of service.

j) Walther PPs for left-handed shooters:

As left-handed shooters usually had problems when handling the safety lever on the left hand side of the slide, Walther/Ulm first of all produced five experimental pistols in October 1976 (SN range: V1048 – V1052) with the safety lever on the right hand side of the slide. So they had the 60° pivoting range for the safety lever milled away on the right hand side of the slide. The pivoting range on the left hand side was still there. The safety shaft was extended to the left to fill the hole in the slide. The safety lever was inversely inserted from the right hand side of the slide.

The right hand side grip plate had to be cut out at the top to make space for the pivoting range of the safety lever.

In the spare part list of Walther/Ulm dated 1982 these slides and safety levers for left-handed shooters are still listed, but these parts were produced only for a very limited period. All in all only three complete "left-handed" PPs were produced by Walther in 1978.

The armorers of the Hamburg police used such parts in the 1970s to modify Walther PPs for left-handed police officers. The serial numbers were subsequently hand-stamped on the slides below the muzzle – mostly with the suffix "L" (for left-handed).

Mainly Walther PPs and hardly any PPKs were modified for left-handed shooters.

k) Walther PPKs with "Rottgardt" Magazine Extensions:

In the 1970s some German police officers reported problems when shooting the Walther PPK, as the grips were too short for big-handed shooters – despite the magazine extensions. From December 1971 onwards a bigger and more voluminous magazine extension designed by G. Rottgardt (then armorer of the Schleswig-Holstein police) solved this problem.

This design used the original floor plate of the PPK magazine, so that the enlarged magazine extension could be fitted to the magazine body. The new magazine extension reached deeper and much further to the rear up to the edge of the grip. At the bottom there was a cutout for the lanyard loop and a hole to handle the floor plate holder.

These hand-casted magazine extensions were first made of a light brown polyester and later of a black polyurethane plastic.

Not only the Schleswig-Holstein police, but also the Hamburg and Bremen police armorers provided such magazine extensions for the needs of their big-handed police officers. Interestingly enough the police in the southern federal states seem to have had smaller hands, as there was obviously no need for such a design there.

l) Walther PPs and PPKs with Engravings:

Quite like with the Zella-Mehlis made PPs and PPKs there were also Walther/Ulm made engraved pistols for sale. As the engraving department of Walther/Zella-Mehlis was very small, there were no factory engraved PPs and PPKs. This is true for the Walther/Ulm made PP/PPK pistols, too, as there was and is no factory engraving department in Ulm at all. So the much quoted "factory engravings" simply do not exist. There were engraved pistols sold by Walther/Ulm, but these engravings were all made by subcontractors – first by Josef Aichholzer, later by Dieter Stübler (both located in Ulm).

m) Walther PP Experimental Weapon 13 rounds:

As the magazine capacity of the Walther PP was criticized by some users, Walther/Ulm experimented in 1984 with a pistol with a double row 13 rounds magazine. There had been a similar experiment by Walther/Zella-Mehlis with three experimental pistols: SN 321 245P, 321 250P and 321 257P. These three pistols were fully automatic pistols, however.

In 1983 Walther/Ulm modified a 9 mm kurz PP made by Manurhin (SN 76 711). The grip was widened by overlay welding to accommodate the double row magazine. The magazine release push button is on the left hand side of the frame and there is an additional slide hold-open lever next to it. The grip plates are made of smooth wood – like with many experimental pistols.

For whatever reasons, this pistol model did not make it to a series production.

n) Walther PP Special Edition "50 Jahre Walther PP" (1979):

To celebrate the 50th anniversary Walther/Ulm issued a limited special series of Walther PPs (800 pistols) in the two calibers 7.65 mm Browning and 9 mm kurz.

SN range:
CW 001 - CW 800

Characteristics: as PPs from the series production, but

- Slide legend on left hand side: "50 Jahre (Walther banner) PP 1929 -1979"
- Slide legend on right hand side: "CW (plus SN) Cal …"
- Inscriptions inlaid in gold
- Carved wooden grip plates
- Trigger, hammer and safety lever gold-plated

o) Walther PPK Special Edition "50 Jahre Walther PPK" (1981):

To celebrate the 50th anniversary Walther/Ulm again issued a limited special series of Walther PPKs (1000 pistols) in the two calibers 7.65 mm Browning and 9 mm kurz.

SN range:
CW 001 - CW 1000

Characteristics: as PPKs from the series production, but

- Slide legend on left hand side: "50 Jahre (Walther banner) PPK 1931 -1981"
- Slide legend on right hand side: "CW (plus SN) Cal …"
- Inscriptions inlaid in gold
- Carved wooden grip plates
- Trigger, hammer and safety lever gold-plated

p) Walther PP/PPK Special Edition "Last Edition" (12/1999):

The last Walther PPs and PPKs to be produced have the additional legend "Last Edition" and "1929 – 1999" resp. "1931 – 1999" on the right hand side of the slide.

q) Walther PPK with incorrect model designation:

In the serial number range around 160 440 A, some PPKs in caliber 9 mm kurz were stamped with "Mod. PP" on the slide (instead of "Mod. PPK").

Walther PPK/S Manurhin and Walther/Ulm

Picture 31: *Walther PPK/S, Walther/Ulm slide legend, SN 167 530S, caliber 9 mm kurz (.380 ACP), produced by Manurhin in 1972*

Production:	1968 - to date (but in caliber .22 l.r. only)
Calibers:	7.65 mm Browning = .32 ACP (8 rounds) 9 mm kurz = .380 ACP (7 rounds) . 22 l.r. (9 rounds)
Total Length:	6.1 in.
Barrel Length:	3.3 in.
Weight:	22.7 oz.
Rifling:	6 grooves, right-twisted
Operation:	unlocked blowback action with firing pin and exposed hammer, double action/ single action trigger mechanism

Characteristics: as Walther/Ulm PP/PPK, but

- Combination of a PP frame with a PPK barrel and slide; this was done to meet the regulations of the new US Gun Control Act of 1968, which prohibited the import of handguns with frame and slide dimensions below certain limits (so-called "Saturday night specials") – e.g. the Walther PPK
- Suffix "S" in PPK/S for "Spezial" (early PPK/S pistols still have this designation spelled "PPKS" – without the slash)
- PPK/S pistols produced under license in the US carry the additional inscription "Under License of Carl Walther Ulm/Do." on pistols made for Interarms, resp. " ... Licence" ..." on those made by S&W – cf. US made PPKs.
- The last variation of these PPK/S pistols, produced to date in caliber .380 ACP by Smith & Wesson, has a longer and

more pointed rear grip tang (so-called "beavertail").

- PPK/S magazines are identical to the ones for the Walther/Ulm PPs

Remarks:

The serial numbers of Manurhin and Walther/Ulm made PPK/S pistols are in the Walther PPK serial range, but with the suffix "S".

In 1983 Walther had some PPK/S frames made by Ranger Manufacturing in Gadsden/ Alabama shipped to Ulm via Interarms. These frames were not forged, but investment casted and some of them already serial numbered by Ranger (number range: W 00xxx). Then the numbers were preset in the mold already. Others had the serial numbers subsequently engraved in the frames. Walther completed these frames with slides, barrel and minor parts made by Manurhin. These pistols were then sold via German retailers as special offer PPK/S pistols.

The Walther PPK/S was given a NATO supply number (1005-21-898-7788) and a military service designation ("Pistole 9 mm Kurz PPK/S"), but was not procured by the German "Bundeswehr".

PPK/S caliber .22 l.r. (Manurhin only!):

steel frame (1969 - 1987):
SN 114 481 – 141 447
Dural frame (1969 - 1982):
SN 507 101 - 507 430
production re-start by Walther/ Arnsberg:
4/2013

PPK/S caliber 7,65 mm Browning (steel frames only):

Manurhin (1969 - 1984):
SN 238 581 - 329 527
Walther (1985 - 1989):
SN 800 001 - 806 271 (about 2000 pistols)

PPK/S caliber 9 mm kurz (steel frames only):

Manurhin (1968 - 1985):
SN 134 941 - 269 008
Walther (1985 - 1989):
SN 800 048 - 804 970 (about 2300 pistols)

Total production of Walther PPK/S Manurhin and Walther/Ulm pistols:
about 115,100 pistols

Walther PPK/S New Edition Walther/Arnsberg

Walther/Ulm never produced any PPK/S pistols in caliber .22 l.r. . All PPK/S pistols made in that caliber were produced by Manurhin. Due to the strong demand for this model – especially in the US – Walther decided to re-start the PPK/S production in caliber .22 l.r. from March 2013 onwards in the Walther subsidiary in Arnsberg.

On the left hand side of the slide is inscribed (although the place of production is not Ulm!): "Carl Walther Ulm/Do. / Modell PPK/S Cal. .22 LR" with the Walther banner on the left hand side of the inscription.

The proof marks are the German ones of the Cologne proof house (shield with three bishops' hats). Cologne is simply closer to Arnsberg than Ulm.

Characteristics: as previous PPK/S pistols, but

- Not made of blued steel or stainless steel, but of die-cast zinc alloy with cast-in steel inserts. The
- Finish: black or a brushed stainless look
- Proof marks: Cologne proof house ("Federal Eagle over N" with "three bishops' hats" and year code)
- Pistols for the US market: "Walther Arms Fort Smith, AR" on right hand side of slide

SN ranges:

pre-series:	from SN PL 001 001
series prod.:	from SN WF 001 001

Walther PPK/S Licensed US Production

Picture 32: *Walther PPK/S, Walther/Ulm slide legend, SN 156 551S, caliber 9 mm kurz (.380 ACP), produced by Smith & Wesson in 1999*

The very first PPK/S pistols for the US importer, Samuel Cummings from Interarms Alexandria/Virginia, from 1968 were made by Manurhin and marked "Made in France". They were first made of blued steel and later of stainless steel.

From 1979 Cummings had these PPK/S pistols produced in the US in caliber 9 mm kurz (.380 ACP) and made of stainless steel only. From 1998 he had them produced in caliber 7.65 mm Browning (.32 ACP), too. On the right hand side of the slide these pistols have the Interarms name and address.

Interarms did not produce these pistols itself, but from 1979 onwards Cummings had them manufactured first by Ranger

Walther PPK/E Walther/Ulm

Picture 33: *Walther PPK/E, Walther/Ulm slide legend, SN 023 034, caliber 9 mm kurz (.380 ACP), produced by FEG in 2003 ("CAT 12654" = customs number for the export to Italy)*

Production:	2000 - 2010
Calibers:	7.65 mm Browning = .32 ACP (8 rounds) 9 mm kurz = .380 ACP (7 rounds) . 22 l.r. (9 rounds)
Total Length:	6.5 in.
Barrel Length:	3.3 in.
Weight:	23.6 - 25.4 oz.
Rifling:	6 grooves, right-twisted
Operation:	unlocked blowback action with firing pin and exposed hammer, double action/ single action trigger mechanism

Characteristics: as Walther/Ulm PP/PPKs, but

- Slide legend on left hand side of slide: "Carl Walther Waffenfabrik Ulm/Do. / Modell PPK/E Cal. …" with Walther banner
- PPK/E pistols were produced with steel frames only.
- Modified firing pin safety: When the trigger is pulled, the firing pin is lifted and only then it can be reached by the hammer – similar to FEG pistols model SMC resp. APK
- Magazines similar to the ones for FEG pistols SMC resp. APK with a slightly bigger capacity than the ones for Walther/ Ulm PPKs – see there.
- Modified grip plates with a thumb rest – similar to the ones for FEG pistols model SMC resp. APK

Remarks:

The Walther pistol PPK/E is dealt with here only, as it is not really a Walther product – despite the slide legend. All PPK/E pistols were produced by the Hungarian company

"FegArmy Kft. Budapest" (short: "FEG"). The design features are mostly identical with the ones of the FEG pistols model SMC resp. APK.

The letter "E" stands for "Europa" and not – as sometimes stated ironically – for "Economy".

With this product Walther wanted to cover the lower price segment, but without much success.

Serial number ranges:

PPK/E 7.65 mm Browning:	from SN 001001
PPK/E 9 mm kurz:	from SN 020001
PPK/E .22 l.r.:	from SN 040001

Total production of Walther PPK/E pistols: about 6,800 pistols

Walther PP and PPK Post-1945 Copies

Preliminary remarks:

When dealing with PP/PPK copies one has to differentiate between copies in the narrow sense (i.e. "clones") and copies in the wider sense, i.e. pistols, which have only some design features in common with the Walther pistols.

You will find a detailed survey of these copies in: Dieter H. Marschall, "Lizensiert und abgekupfert – Walther PP/PPK-Kopien und Nachbauten", in: Deutsches Waffen-Journal, No. 1 – 4, 1998.

The following is only a short summary.

Copies of the Walther PP:

a) ex-GDR ("Pistole 1001"):

In 1947 a team of twelve designers under the supervision of two Soviet officers sat together in the former "Röhm-Werke"/Suhl (later "August-Bebel-Werke") in order to reconstruct old design drawings for a post-1945 production of Walther PPs and other models in the "Soviet Occupation Zone".

From 1953 to 1957 these drawings were used to produce type Walther PP pistols and parts in a secret department of the former "Haenel-Werk" in Bahnhofstraße/Suhl. This secret department was called "Spezielle Produktion" ("special production" – i.e. secret military production). At that point of time "Haenel" was already part of the large, state controlled concern "VEB Ernst-Thälmann-Werke" and the former "Soviet Occupation Zone" had become the "German Democratic Republic" (GDR) – i.e. East Germany.

Some Walther PPs, which were kept in storage, were refurbished and if necessary fitted with new parts. Then again totally new pistols were produced from new parts. New P.38 pistols and parts for these, as well as new P08 pistols were made, too. Newly manufactured pistols were often given serial numbers beginning with the prefix "N" (for "neu") – see "Walther P.38 ex-GDR", too.

As in those days the production of war material on German soil was forbidden by the "Four-Power Agreement", this production was given the code designation "1001" (the code number for "special production"). Accordingly these Walther PP clones were given the designation "Pistole 1001". Depending on the caliber they were marked on the left hand side of the slide with "1001-0-Cal. 7,65", "1001-0-Cal. 22" or "1001-0-Cal. 5,6".

There only a few PP clones were made as experimental pistols in caliber 9 mm Makarov.

Picture 34: *ex-GDR Walther PP copy "Pistole 1001", SN 6090, caliber 7.65 mm Browning (.32 ACP), produced by VEB "Ernst-Thälmann", with "A.B.70" acceptance stamp*

They were marked "1001-3-9,2K". So far only the serial numbers 00014 and 00017 are known, however.

Three experimental pistols were made in caliber 9 mm kurz, too, which were marked "1001-3-9 k". Here only the serial numbers 00001, 00002 and 00011 are known.

These five pistols have a bottom magazine release and a lateral lanyard loop at the heel of the grip. Both characteristics cannot be found on series production "P 1001". So these five specimens are obviously experimental pistols (cf. the extremely low serial numbers, too).

In the late 1950s the "Ministerium für Staatssicherheit" (short: "MfS" or "Stasi") had about 200 "Pistole 1001" produced with the bottom magazine release and in caliber 9 mm kurz. Due to logistical reasons and problems with the acquisition of suitable ammunition these pistols were never issued, however, but went into storage.

PP clones, which were assembled from parts from old stock do not carry the designation "1001- ..." on the left hand side of the slide, but often the old Walther Zella-Mehlis designations still.

Many of the newly produced resp. assembled pistols are marked with "A.B. 79" or "A.B. 70". These are acceptance stamps of the so-called "Abnahmebeauftragte" (acceptance inspectors). The East-German inspection agency is comparable to the West-German "Güteprüfdienst des Bundesverteidigungsministeriums". Up to 1956 (the founding date of the "Nationale Volksarmee", short: "NVA" = National People's Army) these acceptance inspectors reported directly to the Ministry of the Interior. From 1956 onwards the Ministry of National Defense had its own acceptance inspectors. These inspectors sat directly in the factory "VEB Ernst-Thälmann" and inspected weapons for both ministries. They stamped the weapons, which had passed with "A.B. 9" or "A.B. 70". The stamp "A.B. 9" was used by other departments of the ministry, too – e.g. to mark furniture.

If a "Pistole 1001" was made for the national police ("Volkspolizei", short: "VOPO", which was subject to the Ministry of the Interior), it was marked additionally with the so-called "VOPO star" (a stylized police star or badge with a shield in its center). The numbers in the shield designate the VOPO district, where this pistol was issued to – see list in the appendix.

The "Pistole 1001" is identical with the 6th variation Walther PP – see there. It shows the rough machining, the longer grip tang and has no step at the base of the trigger guard. Most of these pistols have a loaded chamber indicator though. The grip plates are made of greenish brown plastic and have no Walther banner.

Some "Pistole 1001" show the designation "Zella-Mehlis Germany" on the left hand side side of the slide – next to the slide legend and also the stamp "ac". These stamps were applied subsequently by US dealers in order to "upgrade" this product. The same is true for the subsequent stamp "Made in Germany" on "Pistole 1001" specimens.

Some "Pistole 1001" have the stamp "Hämmerli Tiengen" inside. In September 1985 the Swiss company Hämmerli purchased 3000 "Pistole 1001" via an Austrian trading company. These pistols were inspected by Hämmerli, stamped and then sold on the commercial market.

The East-German service pistols "Pistole 1001" were taken out of service after 1990, when the two Germanies were reunited.

There were no Walther PPK clones made in East Germany although from 1945 to 1990 many PPKs were used as service pistols by the so-called "armed organs" of the GDR.

If the fragile one-piece PPK grips broke, they were replaced by two-piece ones made either of smooth, oiled beech wood or of black plastic – both without the Walther banner.

In 1994 the company "VEB Ernst-Thälmann" was renamed "Suhler Jagd- und Sportwaffen GmbH" and relocated to the Friedberg area in Suhl.

b) Hungary:

The company "Femaru, Fegyver-es-gepgyar" (short: "FEG"), later renamed "FegArmy" in Budapest produced post-1945 PP copies in the calibers 7.65 mm Browning, 9 mm kurz and 9 mm Makarov. These were sold with the model designations "Walam 48", "PA63", "AP66" or with misleading, invented names like "Welt-Waffen", "Hege" (Hege was a German importer), "Attila" – or even "Luger" and "Mauser Oberaudorf" (The famous German company Mauser is situated in Oberndorf!).

c) Turkey:

From 1947/48 onwards the Turkish company "Makina ve Kimya Endüstrisi" (short: "MKE") in Kirikkale produced pistols of the type Walther PP in the calibers 7.65 mm Browning and 9 mm kurz. These pistols are almost identical with the 4th variation Zella-Mehlis PPs in caliber 9 mm kurz (see there), but have spur shaped magazine extensions (similar to the ones on the Beretta Mod. 34/35). Quite like the 9 mm kurz Zella-Mehlis PPs they have the bottom magazine release and the ejector with an additional "spur".

In 1936 already the Turkish company had obviously ordered a sample pistol for this post-1945 production. A Walther PP pistol with serial number 934 386 was shipped to Turkey which had the following characteristics:

- Turkish acceptance stamp (star with a crescent)
- Slide legend: "T.C Ordusu Subaylarina Mahsustur" (officer's service pistol)

During a trip to Turkey Karl-Heinz Walther and his wife Ira talked with MKE directors about a license contract, but such a contract was never concluded.

Picture 35: *Hungarian Walther PP copy "Walam 48", SN E07823, caliber 9 mm kurz (.380 ACP), produced by FEG Budapest*

Characteristics: as the 4th variation Zella-Mehlis Walther PP, but

- Slide legend on left hand side: "Kirikkale Tüfek Fb. Cap 7,65" (resp. "... Cap 9 mm"); later specimens have the additional company logo "MKE" above this legend.
- Slide legend on right hand side: year of production and with some specimens an additional legend "T.C Ordusu Subaylarina Mahsus" (roughly translated: "for Turkish army officers only")
- Bottom magazine release
- Grip plates: inscribed either with "MKE" or with a banner similar to the Walther one, but with the legend "Kirikkale" inside

d) Czechoslovakia:

The company "Ceska Zbrojovka" (short: "CZ") produced Walther PP copies in the wider sense: model CZ 50 (from 1948) and model CZ 70 (from 1979). Both models have the safety lever on the left hand side of the frame (within the grip plate).

e) Spain:

The pistol model "Constable" made by the Spanish company "Astra" in Guernica-Viscaya (calibers: 7.65 mm Browning, 9 mm kurz and .22 l.r.) is a copy of the Walther PP in the wider sense, too, but with a modified disassembly mechanism and slide hold-open lever.

f) Italy:

The major characteristics of the Beretta pistol model 90 made in Brescia are based on the Walther PP, too.

g) Soviet Union

The Makarov pistol (PM9) which was produced from 1951 in caliber 9 mm Makarov (9 x 18 mm) is based on the Walther PP as well – with a modified safety and firing mechanism. A manual slide hold-open lever was added.

Picture 36: *Chinese Walther PPK copy "Mod 356", SN 016124, caliber 7.65 mm Browning (.32 ACP), with Chinese stamp "jiang" (gun control) on rear grip tang*

Copies of the Walther PPK:

a) People's Republic of China:

The pistol "Model 356" in caliber 7.65 mm Browning is an exact copy of the 6th variation Zella-Mehlis PPK (see there). This model shows the longer grip tang, too, and has no more milled step at the base of the trigger guard.

On the left hand side of the grip tang there is the Chinese character "jian" (for "control"). The slide legend on the left hand side of the slide contains the serial number, the Chinese "Communist Star" and the year of production (mostly 1952).

These pistols are unlicensed copies of the Walther PPK. The different parts are roughly machined only, but can easily be exchanged with Zella-Mehlis made ones, as they are absolutely identical. So one can well talk of a "PPK clone". Even the Walther banner on the grips was imitated, but with a different Chinese inscription inside.

According to unconfirmed rumors these pistols were produced on machines, which were taken from Zella-Mehlis after 1945 and given to the Chinese under political and military assistance agreements.

There is at least one more Chinese PPK clone known with the SN 894 189. This pistol corresponds precisely to the 4th variation Zella-Mehlis PPK (see there). The machining is very rough only and the slide legend is made by hand and not even professionally either. The grips are coarsely carved from a single piece of wood.

There are good reasons to assume that this pistol (and others?) was made from 1958 during the Communist campaign "Great Leap Forward" – at a time, when the Communist regime tried to industrialize the country by force. So almost every village was encouraged to establish its own "metal industry" and tried to fabricate its own metal products – including weapons. It is, however, hardly advisable to shoot such a pistol as the materials and the

Picture 37: *Romanian Walther PPK copy "Md. 74", SN AM-2246, caliber 7.65 mm Browning (.32 ACP), produced by "UMC" in 1980*

workmanship are of inferior quality.

b) Hungary:

The company "Femaru-es-Szersam-gepygar", later renamed "FegArmy Kft." (short: "FEG"), in Budapest produced post-1945 PPK copies in the calibers 7.65 mm Browning, 9 mm kurz and 9 mm Makarov. These were sold with the model designations "SMC" or "APK" – or even "Luger".

c) Romania:

The state-owned company "Usina Mecanica Cugir" (short: "UMC") in Cugir/Transylvania produced a PPK copy (in the wider sense) with a lightweight frame – the "Mod. 74". This pistol was commissioned by the Romanian police and army as service pistol "Md. 74". About 28 000 specimens were shipped to the ex-GDR as well.

As late as 1990 – after the reunification of the two Germanies – some of these pistols from the ex-GDR were issued to some authorities of the "Bundesministerium des Innern" – "BMI" (Ministry of the Interior). These pistols were marked with the property stamp "BMI" and the acceptance stamp of the "Güteprüfdienst des Bundesverteidigungsministeriums".

d) Poland:

The pistol model "P-64" was made by the gun factory "Lucznik" in Radom/Poland. Although there was a Polish patent granted for this pistol, it may well be regarded as a Walther PPK copy in the wider sense.

e) USA:

In the 1970s George Perlotto produced a PPK copy in his company "Indian Arms" in Detroit/Michigan. This pistol was made of stainless steel in caliber .380 ACP – hence the model designation "P.380". There were about 1000 pistols produced – not under license.

Picture 38: *US American Walther PPK copy "P.380", SN 003806, caliber 9 mm kurz (.380 ACP), produced by Indian Arms in 1976*

The company "Indian Arms" was later renamed "Eagle Arms" and then to "Michigan Armament Guardian".

Characteristics: as the Interarms PPKs, but

- Slide in the muzzle area not rounded, but totally square
- Ventilated slide rib
- Two-piece wooden grip plates with a metal insert displaying the company logo (an Indian chief's head)

Walther PP-Super Walther/Ulm

Production:	5/ 1974 - 1981
Calibers:	9 mm Ultra = 9 mm Police = 9 x 18 mm (7 rounds) or rarely: 9 mm kurz = .380 ACP (7 rounds)
Total Length:	6.7 in.
Barrel Length:	3.6 in.
Weight:	26.8 oz.
Rifling:	6 grooves, right-twisted
Operation:	unlocked blowback action with firing pin and exposed hammer, double action/ single action trigger mechanism

Characteristics:

- Slide legend on left hand side: "Carl Walther Waffenfabrik Ulm/Do. / PP-Super Cal. ... / Made in W.-Germany" (with

Picture 39: *Walther PP-Super, Walther/Ulm slide legend, SN 003806, caliber 9 mm kurz (.380 ACP), produced by Walther/Ulm in 1980*

Walther banner)

- Slide legend on right hand side: none
- Proof marks: "Federal eagle over N" with mark for Ulm ("antler")
- SN horizontally on right hand side of slide, with pistols in caliber 9 mm kurz SN additionally on the frame, too
- Fixed barrel, which serves a guide for the recoil spring
- Pivoting, spring-loaded trigger guard, which serves as a dismantling device – cf. Walther PP/PPK; wider trigger guard for shooting with gloves
- Trigger freestanding, protected by trigger guard
- Sights: guilloché slide rib with milled front sight, adjustable rear sight (inserted in a dovetail)
- 26 slide serrations on both sides of the slide
- Ejection port on right side of slide
- Exposed hammer with a serrated spur
- Main spring in the backstrap of the frame
- Internal trigger bar
- Magazine release: push button on upper left hand side of frame (upper magazine release)
- Pivoting de-cocking lever instead of a safety lever (in left hand side of the slide within the serrated area), similar to the Walther P4 – see there
- "automatic" (shooter independent) safety: After de-cocking the hammer the pistol is automatically in the "safe" mode (shooter independent) via a cutout in the hammer: The firing pin is then held down by a spring-loaded push slider. The firing pin is lifted into the reach of the hammer only when the trigger is pulled - similar to the Walther P4 and the FEG APK resp. the Walther PPK/E – see there.
- Swiveling range of de-cocking lever: 60°
- Slide hold-open lever (in left grip plate): activated by the magazine follower, when the magazine is emptied and deactivated

manually
- Metal finish: high-gloss polished finish
- Grips: two-piece grip plates made of checkered brown or black plastic resp. checkered wood, without Walther banner
- Magazine with inserted floor plate or inserted magazine extension

Remarks:

The Walther pistol PP-Super in the powerful caliber 9 x 18 mm was an attempt to provide a service pistol, which meets the needs of the police forces. This model was developed together with S.F. Hübner, a successful German combat shooter.

The first prototype (SN 1 with a Dural frame) was produced in 1971 already. From November 1972 onwards there were tests with experimental pistols. These pistols had the prefix "V" in their serial numbers ("V" for "Versuch", in English "experiment" or "test"). The 44 experimental pistols in caliber 9 mm kurz (.380 ACP) had the serial numbers V1002 – V1198 (not numbered consecutively) and the 53 experimental pistols in 9 mm Ultra had the serial numbers V1012 – V10 238 (not numbered consecutively). There were even tests with two PP-Super pistols in caliber 7.65 mm Browning: V1010 and V1020.

In 1973/4 there were tests with Dural frames for the PP-Super, too.

The only technical change during the production period of the PP-Super was a modification of the firing pin and of the de-cocking mechanism in 1977 (from SN 16 874 onwards).

The PP-Super was produced up to 1979 only, but sold from stocks up to 1983.

PP-Super	9 mm Ultra: 10 001 - 21 101 (total production: 11 101 pistols - some rare specimens with an SN below 10 001 are known)
PP-Super	9 mm kurz: 100 001 - 101 300 (total production: 1300 pistols, 1000 of these went to the USA)

PP-Super for the German police (all in caliber 9 mm Ultra)

Bavarian police (marked "ByP"): 12/1974 - 7/1979 (SN 10 012 – 18 960 (no closed SN block)

- Baden-Württemberg police (marked with "three stacked lions"): advance delivery 2/1974: SN 10 001 – 10 011 (11 pistols); main delivery 12/1974 – 11/1979: SN 10 015 – 18 952 (total: 2345 pistols, but no closed SN block)
- Rheinland-Pfalz police (marked "R-P"): advance delivery 6/1975: SN 10 024 and 10 031 (2 pistols); main delivery 2/1975 – 1/1978: SN 10 248 – 17 513 (total: 2452 pistols, but no closed SN block)
- Schleswig-Holstein (marked with the state coat of arms "two lions with a nettle leaf"): 5/1975 – 2/183 (only 35 pistols, mainly for personal protection officers)
- Bundeskriminalamt (short:BKA, federal bureau of investigation) in Wiesbaden: procured only 25 pistols in 1975 (two of these in caliber 7.65 mm Browning)

All in all the Walther PP-Super was not very popular with German police officers. So from 1975 to 1979 these pistols were gradually taken out of service again. Some of them were subsequently sold as service pistols to forestry officials.

Total production of Walther PP-Super pistols: about 12,400 pistols

Walther TP ("Taschenpistole") Walther/Ulm

Picture 40: *Walther TP, Walther/Ulm slide legend, SN 001 062, caliber 6.35 mm Browning (.25 ACP), produced by Walther/Ulm in 1961*

Production:	1961 - 1971
Calibers:	6.35 mm Browning = .25 ACP (6 rounds) .22l.r. (6 rounds)
Total Length:	5.1 in.
Barrel Length:	2.6 in.
Weight:	12.0 oz.
Rifling:	6 grooves, right-twisted
Operation:	hammerless, unlocked blowback action, striker-fired, no double action trigger mechanism, single action only

Characteristics:

The Walther TP is technically a direct successor to the Walther Mods. 1 and 9.

- Slide legend either on left hand side: "Modell TP Cal. …" and on right hand side: "Carl Walther Waffenfabrik Ulm/Do." (with Walther banner or (from about 9/1967) slide legend only on left hand side "TP Cal. … Ulm/Do." (with Walther banner)
- Proof marks: "Federal Eagle over N" with mark for Ulm ("antler")
- SN vertically on left hand side of frame
- 14 slide serrations on both sides of the slide
- Magazine release: bottom magazine release with a latch in the heel of the grip
- Safety lever on left hand side of slide in front of the serrated area with a 90° pivoting range (The safety blocks the forward movement of the firing pin and disengages the trigger bar.)
- Magazine extension made of black plastic

- Metal finish: salt blued (high-gloss polished finish), nickel-plated finish or engravings optional at an extra charge

Design characteristics similar to the ones of Walther Mod. 9:

- Stepped slide, which embraces the barrel sleeve only partially, open at the top
- Fixed barrel, pressed into the barrel sleeve, which is part of the frame and reaches to the front of the frame
- Recoil spring is held in a separate recess below the barrel
- Sights: sighting groove on top of the slide and milled front sight on top of the barrel sleeve
- Slide dismantling lever at the rear of the slide
- Ejection to the top
- Rear end of firing pin serves as a signal pin
- Grip plates are held by enamel medallions; but are made of black, checkered plastic (with Walther banner in the lower half);

Use of Walther TP Pistols:

In the 1960s and 70s the Walther TP was used as a service pistol by female plain clothes officers and personal protection officers of the BKA (federal bureau of investigation).

Walther TP 6,35 mm Browning:
SN range: 10 001 - 12 253 (11 253 pistols)

Walther TP .22 l.r.:
SN range: 200 011 - 203 427 (3 417 pistols)

Total production of Walther TP pistols:
14,670 pistols

Walther TPH (Taschenpistole Hahn) Walther/Ulm

Production:	1969 - 2000
Calibers:	6.35 mm Browning = .25 ACP (6 rounds) .22l.r. (6 rounds)
Total Length:	5.3 in.
Barrel Length:	2.9 in.
Weight:	11.8 oz.
Rifling:	6 grooves, right-twisted
Operation:	unlocked blowback action with firing pin and semi-exposed hammer, double action/ single action trigger mechanism

Characteristics: design characteristics are similar to the ones of the Walther PP/PPK

- Slide legend on left hand side: "Carl Walther Waffenfabrik Ulm/Do. / Modell TPH Cal. ..." (with Walther banner) or (from about SN 300 000): "TPH Cal. ..."
- Proof marks: "Federal Eagle over N" with mark for Ulm ("antler")
- Serial numbers: from SN 250 001 (7/1969) on right hand side of frame behind the trigger, from about SN 260 000 (1970/71) on right hand side of slide
- Slide made of salt blued steel, closed at the top,
- 17 serrations on both sides of the slide
- Frame made of anodized Dural
- Pivoting, spring-loaded trigger guard, which serves as a dismantling device
- Trigger freestanding, protected by trigger guard
- Sights: guilloché slide rib with milled front sight and adjustable rear sight (inserted in a dovetail)
- Ejection port on right hand side of slide
- Fixed barrel, which serves as a guide for the recoil spring

Picture 41: *Walther TPH, Walther/Ulm slide legend, SN 275 183, caliber 6.35 mm Browning (.25 ACP), produced by Walther/Ulm in 1973, marked with Bavarian police property stamp below the grip plate*

- Main spring in the backstrap of the frame
- Internal trigger bar
- Bottom magazine release
- Pivoting safety lever (in left hand side of slide behind the serrated area) with a pivoting range of 60°
- Safety system: When in the position "safe" the safety shaft covers the rear end of the firing pin and additionally blocks its forward movement; safety disengages the trigger bar from the sear; hammer can only be cocked when the safety is in the position "fire".
- No signal pin
- No hammer block, but hammer safety catch
- No slide hold-open function when magazine is emptied
- Exposed extractor (in ejection port)
- Two-piece grips made of black, checkered plastic (with Walther banner in the lower half), fitted to the frame with a screw
- Magazine extension made of black plastic
- Walther TPH pistols in the calibers .22 l.r. and 6.35 mm Browning were made in the same serial number range.

Remarks:

Before the series production of the Walther TPH there was a pre-series of 80 TPH pistols in 1968 (SN range 001 – 111).

During the whole production period there were only three technical changes:

- 1970 (from SN 252 300): firing pin lock lever and double action lifter fly (trigger tumbler) were modified. Older TPHs were retrofitted using the conversion kit "Einlage TPH 222 46 07".
- 3/1986 (from SN 295 730): modification of the magazine catch by an additional pin
- Late TPH pistols: The trigger guard can only be pivoted downwards, if the magazine was removed before.

Picture 42: *Walther TPH "stainless", licensed US production, SN T 008 687, caliber .22 l.r., produced for Interarms in 1989*

Use of the TPH pistols:

In the 1970s the Walther TPH was often used as a service pistol by female plain clothes officers and other police staff. Then the police property markings were mostly applied underneath the grip plates. A special holster was issued for the police TPH pistols, which could be used as a shoulder or as a belt holster.

In the "hot" 1970s the German "Bundeswehr" was issued a number of TPH pistols in caliber 6.35 mm Browning. Similar to the "Personal Protection Weapons" of the British forces (see there) these pistols could be used by those members of the "Bundeswehr", who were particularly at risk, for the concealed carrying of a personal defense pistol.

Walther TPH 6.35 mm Browning:
SN range: 7/1969 - 5/2000 (SN 250 201 - 304 970 (about 20 050 pistols)

Walther TPH .22 l.r.:
SN range: 7/1969 - 9/1999 (SN 250 001 - 305 137 (about 35 100 pistols)

Walther TPH Special Edition "25 Jahre TPH"

In 1994 Walther issued a limited edition of their TPH in the calibers 6.35 mm Browning (.25 ACP) and .22 l.r.

Characteristics:

- Slide legend on left hand side: "25 Jahre TPH 1969 – 1994" (with Walther banner on left hand side)
- Slide legend on right hand side: "TPH Cal. ...."
- Slide: steel with a grey ("pickled") finish and scroll engravings
- Frame: Dural with a blue-black anodized finish
- Grip plates made of carved wood
- SN on right hand side of slide (beginning

with CW 001)

Total production of Walther TPH pistols: 55,137 pistols

Walther TPH Licensed US Production

From 1979 these pistols were made for Interarms, Alexandria/Virginia first by Ranger Manufacturing Co., then by Black Creek Inc. – both were parts of the company consortium Mid South Industries in Gadsden/Alabama (cf. licensed PPK and PPK/S production).

From 1986 Interarms had the TPH produced in stainless steel and in caliber .22 l.r. (SN prefixed with a "T"). From 1992 these pistols were made in caliber .25 ACP, too (SN prefixed with an "F"). Then in 1994 the US TPH came in blued steel in caliber .22 l.r. (SN prefixed with an "H").

In 1999 the license contract with Interarms expired.

Walther "Militär-Pistole" M.P. I – V Walther/Zella-Mehlis

The name of this model is a statement about its goal orientation: The German military was looking for a successor to the Luger pistol P08.

Walther saw a chance to obtain this very lucrative contract and developed the models "Militär-Pistole" and "Armee-Pistole" – see below. Both models were not produced in series production, but as prototypes in small experimental series only.

The spelling of both model designations with a hyphen is not in accordance with modern orthography, but was taken over out of historical reasons.

Next to Fritz Walther the then chief designer, Fritz Barthelmes, essentially participated in the development of the "Militär-" and the "Armee-Pistole" up to the later Walther P.38. He is even said to have been the real "father of the P.38".

The development of the "Militär-Pistole" (abbreviated to "Mod. M.P." in the slide legend) was performed in several steps from the "M.P. I", an unlocked blowback pistol (basically an enlarged Walther PP), up to the "M.P. V" a locked breech, short recoil system with a pivoting locking block and a freestanding barrel (similar to the later Walther Mod. HP resp. P.38). Within this development the unlocked blowback models had problems similar to the ones of the Walther Mod. 6 with the recoil of the powerful 9 mm Luger ammunition.

It is difficult to define the demarcation lines between the different variations – the more so, as only limited numbers were manufactured and the transitions of the design characteristics were mostly flowing.

The Walther "Militär-Pistole" was never officially commissioned as a service pistol – despite its programmatic name. Only the later Walther P.38 succeeded in being accepted as a service pistol – see there.

Karl Schäfer wrote a very detailed and thorough survey of the M.P. development in his article "Die Armeepistolen", in: Waffen-Digest 1996. His system of distinguishing the major variations is basically adopted here.

Walther Militär-Pistole M.P. I

Production:	about 1930 - 1931
Calibers:	9 mm Luger or 9 mm Steyr
Total Length:	8.0 in.
Weight:	34.2 oz.

Picture 43: *Walther "Militär-Pistole" ("M.P. I"), Walther/Zella-Mehlis slide legend, SN 5008, caliber 9 mm Luger, produced by Walther/Zella-Mehlis in 1930/31, allegedly former property of Fritz Walther*

Rifling:	6 grooves, right-twisted
Operation:	unlocked blowback action with firing pin and exposed hammer, double action/ single action trigger mechanism

Characteristics:

- Slide legend on left hand side: "Waffenfabrik Walther Zella-Mehlis (Thür.) / Walther's Patent Cal. 9 m/m / Mod. M.P." (with Walther banner on left hand side of slide legend)
- Design characteristics basically like the ones of the early Walther PP, but bigger in all dimensions
- Fixed barrel, which serves as a guide for the recoil spring
- Pivoting safety lever with a 90° swiveling range
- Slide serrations like Walther PP
- Upper magazine release
- Swiveling lanyard ring at the heel of the grip
- Metal finish: high-gloss polished
- Some of the minor parts are color case hardened
- Checkered wooden grip plates
- Known SN range: 5004 – 5009

Walther Militär-Pistole M.P. II

Production:	about 1931 - 1932
Caliber:	9 mm Luger
Total Length:	7.8 in.
Weight:	32.1 oz.
Rifling:	6 grooves, right-twisted
Operation:	retarded blowback action with firing pin and exposed hammer, double action/ single action trigger mechanism

Characteristics: as before, but

- Breech with an integrated delay mechanism (retarded blowback) to compensate some of the recoil
- Hammer with a large spur
- Slide hold-open lever on left hand side of frame with a checkered thumb spur
- Bottom magazine release in heel of grip
- Fixed lanyard loop on left hand side of grip near the heel (similar to the one of the Mod. HP – see there)
- Known SNs: 101 and 104

Walther Militär-Pistole M.P. III

Production:	about 1932 - 1934
Calibers:	9 mm Luger and extremely rarely .45 ACP
Total Length:	8.5 in.
Weight:	33.5 oz.
Rifling:	6 grooves, right-twisted
Operation:	locked blowback breech with firing pin and concealed hammer; two pivoting, lateral locking blocks, double action/ single action trigger mechanism

Characteristics: as before, but

- The exterior shape of the Mod. M.P. III is close to that of the Mod. HP and P.38 pistols – see there.
- Freestanding barrel held in a sliding barrel carrier (similar to the later Mod. HP and P.38 pistols)
- Slide open on top, without a bridge at the front, but one lateral reinforcement rib on either side
- Slide serrations from the rear end of the slide to the beginning of the milled area for the safety lever
- Slide hold-open lever serves as a disassembly device, too
- Two recoil springs located on either side of the frame (similar to the later Mod. HP and P.38 pistols)
- Top cover of slide serves as its own retention spring (similar to the later Mod. HP and P.38 pistols)
- Pivoting safety lever with a 60° swiveling range
- Slide serrations extend to the end of the safety lever well
- Loaded chamber indicator pin
- Concealed extractor
- Fixed lanyard loop in a round depression in left hand side grip plate (similar to the later Mod. HP and P.38 pistols)
- Slot for a shoulder stock at the rear grip tang
- Known SN: 1002

Walther Militär-Pistole M.P. IV

Production:	about 1935 - 1936
Calibers:	9 mm Luger and .45 ACP (only SN 1005?)
Total Length:	7.8 in.
Weight:	32.1 oz.
Rifling:	6 grooves, right-twisted
Operation:	locked blowback breech with firing pin and concealed hammer; one pivoting locking block, double action/ single action trigger mechanism

Characteristics: as before, but

- Slide with a bridge over the front, no more reinforcement ribs on either side
- Separate slide hold-open and disassembly levers (similar to the later Mod. HP and P.38 pistols)
- Known SN range: 1003 – 1005

Picture 44: *Walther "Militär-Pistole" ("M.P. V"), Walther/Zella-Mehlis slide legend, SN 017, caliber 9 mm Luger, produced by Walther/Zella-Mehlis in 1936/37*

Walther Militär-Pistole M.P. V

Production:	about 1936 - 1937
Caliber:	9 mm Luger
Total Length:	7.8 in.
Weight:	32.1 oz.
Rifling:	6 grooves, right-twisted
Operation:	locked blowback breech with firing pin and exposed hammer; one pivoting locking block (under the barrel), double action/ single action trigger mechanism

Characteristics: as before, but

- Exposed hammer with a large, round spur, but without a lightening hole in the middle
- Slide serrations from the rear end of the slide to the middle of the milled area for the safety lever
- Wooden grip plates with a checkered surface (SN 015) or a smooth surface (SN 017)
- Known SN range: 015 – 017

Walther "Armee-Pistole" Walther/ Zella-Mehlis

Picture 45: *Walther "Armee-Pistole", Walther/Zella-Mehlis slide legend, SN 010, caliber 9 mm Luger, produced by Walther/Zella-Mehlis in 1935 (?), long barrel and slot for shoulder stock (below rear grip tang)*

With the development of the "Armee-Pistole" Walther pursued the same purpose as with the "Militär-Pistole" – to obtain the contract for a new military service pistol.

As with the "Militär-Pistole" there was no series production, but only a small series of about 100 experimental pistols was manufactured with flowing transitions of the design features. Some examples have a black anodized lightweight frame. Shoulder holster stocks similar to the ones for the Mauser C96 ("Broomhandle") were made, too. Some "Armee-" and "Militär-Pistole" models show common or at least similar design characteristics.

Whereas the model designation of the "Militär-Pistole" is abbreviated to "M.P." in the slide legend, this is not the case with the "Armee-Pistole". Here the slide legend gives the designation in full form (with a hyphen). In the Walther literature, however, this model designation is often abbreviated to "Walther AP".

Walther Armee-Pistole

Production:	about 1935 - 1937
Calibers:	9 mm Luger or .45 ACP
Total Length:	8.5 in.
Weight:	33.5 oz.
Rifling:	6 grooves, right-twisted
Operation:	locked blowback breech with firing pin and concealed hammer; one pivoting locking block (under the barrel), double action/ single action trigger mechanism

Picture 46: *Walther "Armee-Pistole", Walther/Zella-Mehlis slide legend, SN 028, caliber 9 mm Luger, produced by Walther/Zella-Mehlis in 1936 (?), standard barrel and no slot for shoulder stock (below rear grip tang)*

Characteristics: as before, but

- Slide legend on left hand side: "Waffenfabrik Walther Zella-Mehlis (Thür.) / Armee-Pistole Cal. 9 m/m" (with Walther banner on left hand side of slide legend)
- Slide legend on right hand side: "Walther's Patent"
- The exterior shape of the Mod. M.P. III is close to that of the Mod. HP and P.38 pistols – see there.
- Freestanding barrel held in a sliding barrel carrier (similar to the later Mod. HP and P.38 pistols)
- Slide open on top, with a bridge at the front and one lateral reinforcement rib on either side (located in front half of slide)
- Separate slide hold-open lever and disassembly lever (both with checkered thumb spurs on top)
- Two recoil springs located on either side of the frame (similar to the later Mod. HP and P.38 pistols)
- Top cover of slide serves as its own retention spring (similar to the later Mod. HP and P.38 pistols)
- Pivoting safety lever with a 60° swiveling range
- Slide serrations from the rear end of the slide to the beginning of the milled area for the safety lever
- Loaded chamber indicator pin
- Concealed extractor
- Bottom magazine release in heel of grip
- Fixed lanyard loop in a round depression in left hand side grip plate (similar to the later Mod. HP and P.38 pistols)
- Some examples with slot for a shoulder stock below the rear grip tang
- Different barrel lengths: e.g. SN 029 with 3.0 in. barrel (subsequently shortened?), SN 028 with 5.0 in. barrel, SN 010 with 8.0 in. barrel
- SN 045: slide legends subsequently milled out on both sides
- Known SNs: 02 - 079

Walther Mod. HP ("Heerespistole") and Mod. P.38 Walther/Zella-Mehlis

Preliminary remarks:

The following comments on the Walther pistols Mod. HP, Mod. P.38 and P.38 basically follow the ones made by Warren Buxton in his very well researched and detailed volumes 1 and 2 of "The P.38 Pistols" (1978 resp. 1984), as well as the ones made by Gene Gangarossa in "The P.38 Automatic Pistol" (1993) – see bibliography. In volume 3 Warren Buxton later has revised and enlarged some of the information from the first two volumes. These alterations are taken over here, too.

During 1990 - 1998 Warren Buxton has revised his data again – in cooperation with Orvel Reichert (USA) and Alexander Krutzek (Germany). These alterations were taken over here as well – e.g. as they were published by A. Krutzek in "Deutsches Waffen-Journal" articles (see bibliography).

It is important to bear in mind that the production data and the demarcation lines between serial number ranges are not as distinct as the following comments might make one believe. Mostly they are based on interpolations and estimates. On top of that there were overlaps, when older parts were used in production to use up old stocks. Sometimes old and newly manufactured parts were used simultaneously in parallel production lines.

As to pistols, which were produced for the military, one has to consider the fact that the number of pistols, which were manufactured, is not identical with the number of pistols, which were issued to the military. The amount of pistols, which were rejected by the "Waffenamt" inspectors, is substantial. The eject rate varied, however. It differed as well depending on the manufacturer – Walther, Mauser or Spreewerk.

The variations designated Mod. HP ("Heerespistole" – Mind the spelling without a hyphen!) and Mod P.38 both are the commercial variations of the military service pistol Walther P.38. Any developments and technical changes in the series production were taken over almost simultaneously for all three variations. Unless noted otherwise in the text the caliber is 9 mm Luger.

The Mod. HP and Mod. P.38 pistols were basically made for the commercial market (target shooting, factory securities, etc.), but somehow some of these pistols made it to become service pistols later – e.g. when an army officer privately bought his side arm or when the "Wehrmacht" procured batches of P.38 type pistols.

The Mod. HP and Mod. P.38 pistols differ from the P.38 service pistols for the police or military forces in their slide legends and markings.

When dating Mod. HP and Mod. P.38 pistols it is interesting to consider the relevant patents again. Some design characteristics can be found with "Militär-Pistole" or "Armee-Pistole" already, but there they were mostly not protected by patents yet. The patents for Mod. HP and Mod. P.38 pistols were of course later relevant for P.38 pistols, too.

It is striking that most of the patents referring to the P.38 design characteristics were filed in 1936 already, but granted mostly much later only. In pre-war times patents presumably were no longer of great importance in the arms industry.

- Patent number 664 926: "Aus flachem Blech hergestelltes Pistolenmagazin" ["pistol magazine made of sheet metal", here: box magazine made of folded and spot welded sheet metal] - filed on 20th November 1936, granted on 8th September

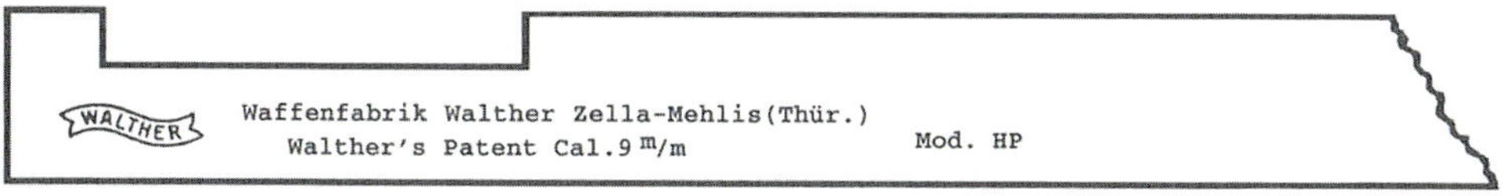

Illustration 4: *example of a Mod. HP slide legend (not true to scale)*

1938

- Patent number 677 094: "Selbstladepistole" ["self-loading pistol", here: breech with two lateral recoil springs] – filed on 14th February 1936, granted on 25th May 1936
- Patent number 678 067: "Entspannsicherung für Selbstladepistolen" ["safety with de-cocking function for self-loading pistols"] - filed on 11th July 1936, granted on 15th June 1939
- Patent number 706 038: "Selbstladepistole" ["self-loading pistol", here: slide with a cover plate] – filed on 10th April 1936, granted on 10th April 1941 only
- Patent number 715 176: "Schlagbolzensperreinrichtung" ["hammer block"] – filed on 2nd April 1938, granted on 20th November 1940
- Patent number 721 702: "Selbstladepistole mit gleitendem Lauf und verriegeltem Verschluß" ["self-loading pistol with a gliding barrel and a locked action"] – filed on 27th October 1936, granted only on 7th May 1942 to Fritz Walther and Fritz Barthelmes
- Patent number 722 332: "Selbstladewaffe ... mit einer Entspannungssicherung" ["self-loading weapon ... with a de-cocking safety"] – filed on 18th February 1939, granted on 21st May 1942 to Fritz Walther and Walter Haasenstein
- Patent number 726 501: "Schloß für automatische Handfeuerwaffen" ["breech for automatic handguns", here: improvements of sear and hammer] – filed on 11th January 1938, granted on 3rd September 1942 to Fritz Walther

The serial numbers of Mod. HP and Mod. P.38 pistols are consecutive in the same number range from SN 1001 to about 25 800. Some Mod. HP pistols have a letter "H" prefix – see "Swedish contract".

Early Mod. HP and Mod. P.38 pistols carry the SN on the right hand side of the frame and on the inside (!) of the slide, on the front of the barrel carrier and on the locking block (here only the last three digits).

Later Mod. HPs and Mod. P.38s (from about SN 7200 onwards) have the SN on the right hand side of the frame, on the top of the slide (on the right rail between the breech face and the bridge – visible only if the slide is retracted), on the front of the barrel carrier and on the locking block (here only the last three digits).

With some pistols the SN may also be located on the left hand side of the magazine body.

Walther Mod. HP and Mod. P.38

Production:	about 1937/38 - 1945
Calibers:	9 mm Luger - 8 rounds 7.65 mm Parabellum = .30 Luger (from 1939, but only very few specimens) - 8 rounds .45 ACP (only experimental pistols) .38 Super (only experimental pistols) . 22 l.r. (only one example known)
Total Length:	8.5 in.

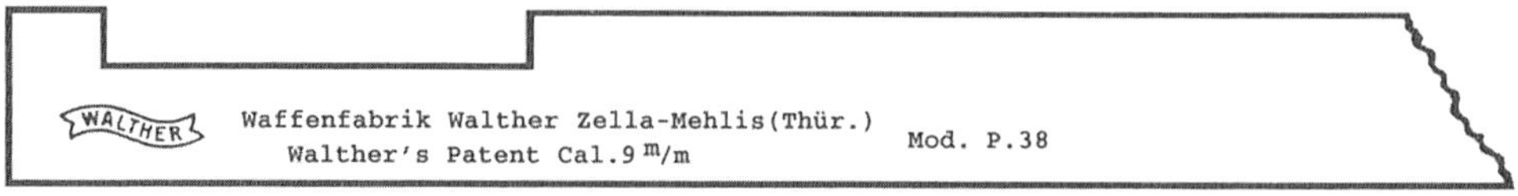

Illustration 5: *example of a Mod. P.38 slide legend (not true to scale)*

Barrel Length:	4.9 in.
Weight:	33.4 oz. (steel frame) or 27.5 lightweight alloy frame
Rifling:	6 grooves, right-twisted
Operation:	locked blowback breech with firing pin and exposed hammer; pivoting locking block (below the barrel), double action/ single action trigger mechanism

Characteristics:

- Slide legend on left hand side: "Waffenfabrik Walther Zella-Mehlis (Thür.) / Walther's Patent Cal. 9 m/m / Mod. HP resp. Mod. P.38" (with Walther banner on left hand side of slide legend)
- Proof marks: from 1937 "Crown over N", from the beginning of 1940 "Eagle over N"
- Frame unreinforced, i.e. frame is flat just below the trigger axle – see Picture 50
- Freestanding barrel held in a sliding barrel carrier, barrel (with its rifling) and barrel carrier are a unit made of one single forged block of steel – i.e. without a pressed-in liner (cf. later P38 Walther/Ulm)
- Barrel carrier slides in one rail on either side of the frame and in two rails on either side of the slide
- Slide open on top, with a bridge at the front, without lateral reinforcement ribs
- Slide with a secondary cutout behind the extractor, i.e. a long extractor relief slot – see Picture 52
- I.e. a long relief slot
- Slide hold-open lever (on left hand side of frame) is activated by the magazine follower, when
- The magazine is emptied
- Pivoting take down lever (on left hand side of frame) serves as a disassembly device
- Two recoil springs located on either side of the frame
- Top cover of slide serves as its own retention spring
- Pivoting safety lever with a 60° swiveling range
- "automatic" (shooter independent) safety: The firing pin is locked in place by a spring-loaded firing pin lock plunger. The latter is pushed upwards (out of contact with the firing pin) by a lever, when the trigger is pulled. shooter dependent (manual) safety: When the safety lever is rotated to the "safe" position, the end of the firing pin is locked in place by a cutout in the safety shaft. The hammer cannot be cocked until the safety lever in the position "fire", as a cam in the safety shaft presses the hammer drop lever down.
- 14 slide serrations on either side of slide (extending to the middle of the safety lever well)
- Loaded chamber indicator pin
- Concealed extractor
- Bottom magazine release in heel of grip
- Fixed lanyard loop in a round depression in left hand side grip plate
- Early grip are plates made of checkered black plastic (phenol resin) or of checkered wood with no serial numbers inside. Later grip plates are similar to the ones for P.38 pistols of the corresponding years of

Picture 47: *Walther Mod. HP, Walther/Zella-Mehlis slide legend, SN 2304, caliber 9 mm Luger, produced by Walther/Zella-Mehlis in April 1940, from second delivery to Sweden (SN without prefix "H", checkered grip plates with small depression for lanyard loop)*

production.

- Metal finish: first high-gloss polished finish, later dull "military" finish

Mod. HP, 1st Variation (1938 – 11/1939)

SN 1001 to 1019 (19 pistols)

Characteristics: as before, but

- Rectangular end of firing pin
- Concealed extractor
- Checkered plastic grip plates, left hand side grip plate with a small, round depression for lanyard loop
- Lanyard ring of the small, round type
- Proof marks: "Crown over N"
- High-gloss polished finish
- Some specimens with a longer, rounded grip tang (for target shooting?)

Mod. P.38, 1st Variation (12/1939 – 1/1940)

SN 1020 to 1067 (48 pistols)

Characteristics: as before, but

- Caliber mostly 7.65 mm Parabellum = .30 Luger
- Some specimens with "single action only" trigger mechanism

Mod. HP for Sweden 1st Variation (1/1940 – 4/1940)

SN H1068 to H2101 (1034 pistols - 999 pistols out of this lot for Sweden)

Characteristics: as Mod. HP 1st variation, but

- SN prefixed with letter "H" (SN on right hand side of frame)
- Exposed extractor
- Take down lever with a checkered thumb

Picture 48: *Walther Mod. HP, Walther/Zella-Mehlis slide legend, SN 20 062, caliber 9 mm Luger, produced by Walther/Zella-Mehlis, horizontally grooved grip plates with rectangular depression for lanyard loop*

spur

- Two magazines with SN on left hand side of body, without acceptance stamps
- Some specimens subsequently stamped with Swedish unit markings (e.g. "I 17 No. 8")

Remarks:

In his new book "A pistol for Sweden – the m/39" (to be published in 2015 – see bibliography) the Norwegian author Per Mathisen explains that the two shipments of Walther HP pistols to Sweden were not based on a contract between the Swedish government and Walther. This business was handled by the Swedish company Widforss/ Stockholm.

The first shipment on 6th April 1940 contained 999 HP pistols (SNs prefixed with "H") and one (!) P.38 Zero Series pistol – a total of 1000 pistols.

The second shipment (see below) on 16th January 1941 consisted of 500 HP pistols (SNs not prefixed with "H").

Both shipments came with the standard Walther boxes, but only the first shipment had serial numbered magazines.

The Swedish armed forces introduced the Walther HP pistols with the designation "m/39".

Mod. HP and Mod. P.38, 2nd Variation (4/1940)

SN 2102 to 2169 (68 pistols)

Characteristics: as before, but

- Mod. HP pistols in caliber 9 mm Luger
- Mod. P.38 pistols in caliber 7.65 mm Parabellum = .30 Luger

Mod. HP, for Sweden 2nd Variation (4/1940)

SN 2170 to 2838 (669 pistols - 500 pistols out of this lot for Sweden)

Characteristics: as Swedish 1st variation Mod. HP, but

- SN no longer prefixed with letter "H"
- Round end of firing pin
- Take down lever with a grooved thumb spur
- Magazine without a SN on body
- Proof marks: "Eagle over N"

Mod. HP and Mod. P.38, still 2nd Variation (4/1940 – 8/1940)

SN 2839 to 3000 (162 pistols)

Characteristics: as before, but

- Some Mod. HP in caliber 7.65 mm Parabellum = .30 Luger, too

Mod. HP, 3rd Variation (9/1940)

SN 3001 to 3100 (100 pistols)

Characteristics: as before, but

- Grip plates no longer made of checkered, black plastic with a round depression for the lanyard loop (on left hand side), but made of horizontally grooved, mottled brown to black plastic with a rectangular depression for the lanyard loop; remark: However sometimes the old type grip plates were still fitted up to SN 5500.
- Lanyard loop no longer small and almost round, but bigger and flatter on top

Mod. HP and Mod. P.38, still 3rd Variation (9/1940 – 4/1941)

SN 3101 to 12 395 (9 295 pistols)

Characteristics: as before, but different modifications of both models, e.g.

- Lightweight alloy frames: about 100 pistols in SN range 6870 – 6932 (11/1940 – 12/1940)
- Wooden grip plates (mostly on pistols in caliber 7.65 mm Parabellum = .30 Luger)
- Some pistols with the marking "Privat" (presumably samples for sales people)
- Some pistols in the calibers .45 ACP and .38 Super (presumably experimental pistols only)
- Some pistols with "Waffenamt" acceptance stamps (only one stamp on right hand side of slide plus commercial proof marks): These pistols were either procured by the German "Wehrmacht" and proofed by the "Waffenamt" inspectors or shipped to an allied state and proofed for them by the "Waffenamt" inspectors in assistance.

Mod. HP, "Croatian Ustasha Contract" (5/1941 – 6/1941)

SN 12 396 to 12 534 (about 100 pistols, but not with consecutive SNs)

Characteristics: as standard 3rd variation Mod. HP, but

- Slide legend on right hand side: letter "U" over a checkered coat of arms ("U" stands for "Ustasha", the coat of arms is the one of the independent state of Croatia founded in April 1941)
- No military acceptance stamps

Remarks:

In May/June 1941 about 100 officers of the fascist Croatian "Ustasha" movement were secretly trained by the German SS to serve as bodyguards for the leader of this fascist movement, Ante Pavelic.

Presumably these officers were equipped by the SS with these Mod. HP pistols. This theory still needs evidence though.

Mod. HP, "Swiss Contract" (1941)

SN range unknown

Only very few Mod. HP pistols were shipped to Switzerland for test purposes.

Mod. HP and Mod. P.38, still 3rd Variation (6/1941 – 2/1945)

SN 12 535 to about 25 800 (about 13 270 pistols)

Characteristics: as before, but several modifications of both models, e.g.

- From about SN 14 000 most pistols with dull "military" finish, as the polishing before the blueing process was omitted (from 1941/42 by order of the "Reichsminister für Bewaffnung und Munition")
- "mixed types": assembled from older and new parts of both models
- Military acceptance stamps or commercial proof marks
- From about mid-1943 the Mod. HP and Mod. P.38 pistols have the reinforced frame and the slide without the secondary cutout behind the extractor – like the P.38 pistols (cf. illustrations 50 – 53).

End of Mod. HP and Mod. P.38 Production Walther/Zella-Mehlis

In February 1945 the production of both models ceased, as exclusively P.38 service pistols were produced for the armed forces from then onwards.

Later some "single action only" pistols were still assembled with early unreinforced frames and longer, rounded rear grip tangs – for the so-called "Volkssturm" ("People's Army") or perhaps for target shooting even?

Total production of Walther/Zella-Mehlis pistols Mod. HP and Mod. P.38: about 24,800 pistols

Post-1945 Use of Walther Mod. HP Pistols

Walther Mod. HP pistols, i.e. war surplus, were reused in post-World War Two days in France, East Germany and in Austria. The latter with the "Austrian Federal Eagle" and the marking "BH" (for "Bundesheer" – see there).

New Edition of Mod. HP Pistols by Walther/Ulm

From July 1998 Walther/Ulm produced a limited series of Mod. HP pistols with a steel frame. This new edition was limited to 100 pistols. They were sold in wooden presentation cases with a spare magazine and a cleaning rod.

Characteristics: as the Walther/Ulm P38s, but

- Slide legend on left hand side (as early Zella-Mehlis Mod. HPs): "Waffenfabrik Walther Zella-Mehlis (Thür.) / Walther's Patent Cal. 9 m/m / Mod. HP" (with Walther banner on left hand side of slide legend)
- Slide legend on right hand side: "1938 – 1998, 60 Jahre Walther Heerespistole, Carl Walther, Ulm/Do. / Made in Germany"
- Metal finish: frame and slide high-gloss polished finish

SN range: HP001 - HP100

Walther P.38 Walther/Zella-Mehlis Production

Unofficially the German "Wehrmacht" (armed forces) accepted the Walther P.38 as a service pistol to succeed the Luger pistol P08 in 1938 already – hence the model designation "P.38".

In 1939 there was a troop trial with 1000 P.38 pistols from the pre-production series (so-called "Nullserie" – "Zero Series").

As a precautionary measure Walther had enlarged its production facilities from August 1939 to produce the new P.38. In that month they manufactured a small series of "Zero Series" P.38s already. By the end of 1939 a total of 1470 P.38s were produced.

The formal adoption as a military service pistol was made on 20th April 1940 (Adolf Hitler's 51st birthday!) by recognizing the 3rd variation "Zero Series" P.38 (see there) as the future standard military pistol.

As late as July/August 1940 did Walther start the full series production of P.38 pistols for the "Wehrmacht" – with a reject rate of about 40% to start with.

The standard caliber of the P.38 pistols is 9 mm Luger. There is only one prototype in caliber .22 l.r. known – without a serial number.

Walther P.38 Grip Plates

The checkered black grip plates for the Mod. HP, Mod. P.38 and "Zero Series" pistols were made by the Walther in-house plastic casting department. Presumably Walther later had P.38 grip plates produced in a branch factory in Wasungen near Meiningen/Thuringia, a former paper mill. This factory was bought by Walther in 1936 (cf. production of PP/PPK grip plates – see above).

These early grip plates are marked inside with four circles, which contain (from top to bottom): the early Walther logo ("Cewe" – for Carl Walther?), the number "480" (an early code number or part designation?), the logo of the "Materialprüfungsamt Berlin-Dahlem" – "MPBD" (in English: material testing authority – see appendix) with the code for the producer "V7" (Walther), the type designation of the material ("Z3" = pulp as a carrier for resin or "T1" = textile fibers as a carrier) and finally the number "1" (position in the casting mold).

Grip plates for the "Zero Series" P.38s additionally have the last three digits of the serial number and the "Waffenamt" inspection stamp ("Eagle/359") inside. These two markings are found on all subsequent Walther P.38 grip plates up to P.38 ac42 pistols.

Towards the end of the "Zero Series" the grip plates show the later standard features, i.e. 31 horizontal grooves on either side of the grip. Out of these 31 grooves six are interrupted near the grip screw – cf. grip plates made for Spreewerk P.38s. The depression for the lanyard loop is no longer small and round, but bigger and rectangular.

These standard grip plates were either made by Walther themselves or later provided by AEG in Henningsdorf near Berlin. They show three circles inside containing (from top to bottom)…

Walther grip plates – the MPBD logo with: "V7" (code for Walther) and "57", "41", "31" or "Z3" (code number for the material with additives); the two lower circles are empty.

AEG grip plates plates – the MPBD logo with: "38" (code for AEG) and "Z3" (code designation for the material with additives); the middle circle is empty; the lower circle with "P 1528" and "P1529" (part numbers for left hand side and right hand side grip plates) plus a number from 1 to 9 (designating the position in the casting mold).

From mid 1944 some grip plates were made of "Durofol", too. This is a laminated material made of resin and pressed wood produced by Brangs & Co., Solingen. The name "Durofol" (in script and embedded in a rhombus) is located inside the grip plates.

Walther P.38 Magazines

Magazines for the Walther made P.38s were either made by Walther themselves or (from 1943) provided by "Erste Nordböhmische Metallwarenfabrik" (First North Bohemian metal wares factory) in Niedereinsiedel/Sudetenland (code stamp "jvd", acceptance stamp "Eagle over WaA 706")

Remarks:

You can find a very detailed description of the different variations of P.38 grip plates and magazines in: Dennis de Vlieger, Ron Clarin and Wolf-Dietrich Roth, "P.38 Magazine und Griffschalen", Borsdorf 2011.

This research was made difficult by the fact that magazines and grip plates were often interchanged during the service life of a P.38 pistol.

Proof and Acceptance of Walther made P.38 Pistols by "Heereswaffenamt" ("WaA 359")

Whilst still in-the-white the major pistol parts were stamped with the first "Waffenamt" (army ordnance office) acceptance stamp ("Eagle over 359" – see appendix). This stamp was applied to the right hand side of the slide and the left hand side of the frame. After the blueing and the final assembly the proof firing was made and then the final proof mark ("Eagle over Swastika") was stamped on the right hand side of the slide. After the final inspection by the "Waffenamt" officers a second acceptance stamp was applied to the right hand side of the slide. As a rule it was the same as the first one.

These "Waffenamt" stamps were first applied to the major and most minor parts of the P.38 pistols. From June 1942 up to the end of production the stamping of minor parts was increasingly omitted. Then only the slide (two acceptance stamps) and other parts like frame, barrel carrier, locking block and magazine (one acceptance stamp each) were marked.

Commercial Mod. HP and Mod. P.38 pistols, which were procured by the German "Wehrmacht" have only one "Waffenamt" acceptance stamp, as they already had the commercial proof marks before the purchase. These pistols passed the final "Waffenamt" inspection only.

It is strange that in the acceptance procedure of the P.38 pistols there was no change in the use of "Waffenamt" acceptance stamps from "Eagle over 359" first to "Eagle over WaA 359" later – as it was the case with Walther PPs and PPKs, which were procured by the "Wehrmacht". There only a few P.38s known with at least one "Eagle over WaA 359" stamp. This was probably due to the use of a wrong die.

The exact inner bore diameter was stamped vertically on the left barrel carrier rail – e.g. "8.80" for 8.80 mm or "8.95" for 8.95 mm).

Walther In-house-inspection and Maintenance Marks

Next to the military marks there may be other stamps applied by the factory employees during manufacture. The exact meaning of most of these stamps is still unknown. Some of them are probably only repair or maintenance marks. These stamps may be in the form of letters, numbers, geometric figures (e.g. circle, triangle, oval), six pointed stars or other stars.

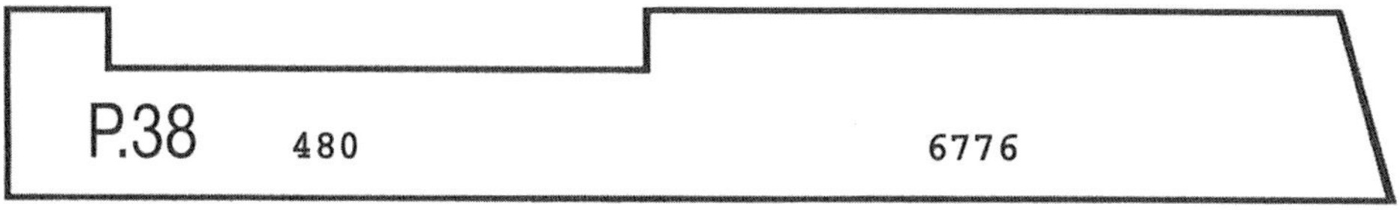

Illustration 6: *example of a Walther P.38 slide legend with code "480"*

A special position is taken here by the so-called "asterisk" (a little star with a hole in the middle, which is similar to the Germanic "hagal rune"):

One theory states that this stamp denotes parts, which were found not to be dimensionally accurate or badly finished, but repairable.

Another theory says that this stamp was used by inspectors of the SS in concentration camps to mark minor pistol parts, which were produced in the camps.

Recent research – especially by Jan Balcar from the Czech Republic – has shown that the first theory is right: Parts marked with the "asterisk" were rejected, refurbished and then used in series production.

One more marking seems to be deciphered: the capital letter "S" (either stamped or electro-penciled). According to the research of the US author Dave Shike this "S" (for "sekundär", in English "secondary") was used to denote parts, which were actually rejected, but used later when pistols for the "Volkssturm" ("People's Army") were assembled as the "last means". According to Dave Shike this assembling of pistols was done from 1942 in "Metallwerke Neuengamme GmbH Carl Walther" a branch factory of Walther/Zella-Mehlis in the concentration camp Neuengamme south-east of Hamburg. There rejected parts, parts from cannibalized pistols and even handmade pistol parts were marked with "S" and used in assembling whole pistols. Fritz Walther was appointed to be the "Wehrwirtschaftsführer" (leader of the military economy) and president of the "Arbeitsgemeinschaft Volksgewehr" (working group people's rifle). As such he had the right to take recourse to gun parts resp. spare parts, which were the property of private people, companies, armed and police forces etc. to have them assembled to complete guns. So this theory seems verified.

Walther P.38 Serial Number System

For the P.38 pistols Walther developed a special serial number system, which differs from the ones for earlier pistols. It is not consecutive, but out of tactical reasons it is deceptive and misleading:

- The P3.8 pistols of the first "Zero Series" were numbered consecutively from "01".
- The numbering of P.38s with code designation began with "1" up to "9999", then went back to "1" suffixed with a letter (beginning with "a" through the alphabet) – i.e. "1a" to "9999a", "1b" to "9999b", etc.
- The blocks "1 – 9999" were, however, not necessarily maintained in a strict sequential order. Sometimes, e.g. at the turn of the year or after changes in production the numbering carried on with a new "1" and the next letter suffix.
- The very first and the very last Walther made P.38 Pistols had serial numbers prefixed with "0" – the "Zero Series".
- As Mauser and Spreewerk introduced a similar numbering system, the same serial numbers could have been issued up to three times – but with different manufacturer code designations.

- Accidentally the numbering machines sometimes skipped a number (e.g. from "9999" to "0000"). Then a "1" was stamped by hand in front of the serial number – to make it "10000".
- According to the US researcher Orvel Reichert thus the two Mauser made P.38s "10000o" and "10000g" (both with the Mauser manufacturer code designation "svw") and the Spreewerk made P.38 "10000" (with the Spreewerk manufacturer code designation "cyq") can be explained.

P.38, Zero Series 1st Variation (8/1939 – 11/1939)

SN 01 - 035 (35 pistols)

Characteristics:

- SNs beginning with "0" (SN horizontally on left hand side of frame and slide)
- Slide legend on left hand side: "P.38" (with Walther banner)
- Proof marks: "Crown over N"
- "Waffenamt" acceptance stamp "Eagle over 359"
- Last three digits of SN even on minor parts
- Frame unreinforced, i.e. frame is flat just below the trigger axle – see Picture 50
- Freestanding barrel held in a sliding barrel carrier, barrel (with its rifling) and barrel carrier are a unit made of one single forged block of steel – i.e. without a pressed-in liner (cf. later P38 Walther/Ulm)
- Barrel carrier slides in one rail on either side of the frame and in two rails on either side of the slide
- Up to about SN 011: ring shaped, fluted chamber to prevent split cartridge cases and the accidental firing of a cartridge due to overheating (fluted chambers so far were known mostly from machine guns)
- Massive locking block with a notch at the front for the wire spring, which holds the block
- Slide open on top, with a bridge at the front, without lateral reinforcement ribs
- Slide with two separately milled out areas at the rear for the signal pin and the hammer
- Slide with a secondary cutout behind the extractor, i.e. a long extractor relief slot – see Picture 52
- Slide hold-open lever (on left hand side of frame) is activated by the magazine follower, when
- The magazine is emptied
- Rectangular end of firing pin – cf. early Mod. HP and Mod. P.38 pistols
- Pivoting take down lever (on left hand side of frame) serves as a disassembly device
- Two recoil springs located on either side of the frame
- Top cover of slide serves as its own retention spring
- Pivoting safety lever with a 60° swiveling range
- Shooter dependent (manual) safety: When the safety lever is on "safe" the firing pin recedes out of the reach of the hammer.
- 14 slide serrations on either side of slide (extending to the middle of the safety lever well)
- Hammer with eight cocking grooves on top
- Loaded chamber indicator pin
- Concealed extractor
- Bottom magazine release in heel of grip
- Grip plates made of black, checkered plastic with a small, round depression on left hand side grip plate for the lanyard loop
- Lanyard loop in the small, round shape
- Metal finish: first high-gloss polished finish
- Magazines with SN (on left hand side of body) and acceptance stamp "Eagle over 359" (on spine), but without model designation "P.38"

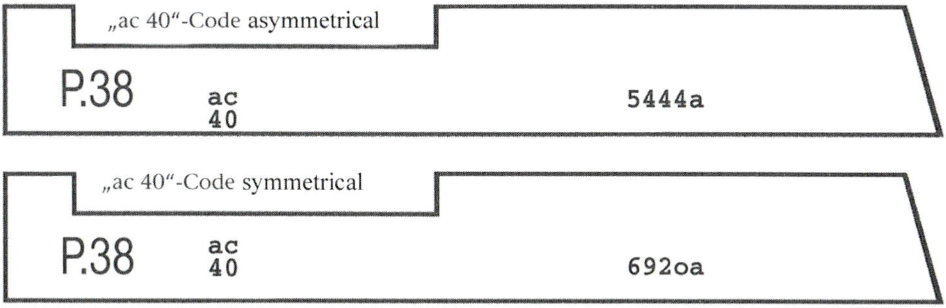

Illustration 7: *examples of Walther P.38 slide legends with code "ac": date "40" added asymmetrically by hand-stamp (above) and date "40" integrated symmetrically in roll-stamp (below)*

P.38, Zero Series 2nd Variation (11/1939 – 1/1940)

SN 036 - 01 000 (about 960 pistols)

Characteristics: as before, but

- No more "receding" firing pin, but "automatic" (shooter independent) safety: The firing pin is locked in place by a spring-loaded firing pin lock plunger. The latter is pushed upwards (out of contact with the firing pin) by a lever, when the trigger is pulled; shooter dependent (manual) safety: When the safety lever is rotated to the "safe" position, the end of the firing pin is locked in place by a cutout in the safety shaft. The hammer cannot be cocked until the safety lever is in the position "fire", as a cam in the safety shaft presses the hammer drop lever down.

P.38, Zero Series 3rd Variation (1/1940 – 4/1940)

SN 01 001 - 03 500 (about 2500 pistols)

Characteristics: as before, but

- Exposed extractor
- From about SN 02 000: two "Waffenamt" acceptance stamps on spine of magazine
- In this 3rd variation the P.38 was accepted as a military service pistol on 20th April 1940.

P.38, Zero Series 4th Variation (4/1940 – 7/1940)

SN 03 501 - 013 720 (about 10 200 pistols)

Characteristics: as before, but

- Round end of firing pin
- No longer separately milled out areas for hammer and signal pin at the rear of the slide, but both millings are merged
- Front of locking block no longer with a notch, as from now onwards held by a plate spring (instead of a wire spring)
- From about SN 012 000 (mid-1940): checkered plastic grip plates with the small, round depression for the smaller, rounded lanyard loop on the left hand side. grip plate now gradually replaced by such made of horizontally grooved, mottled brown to black plastic with a rectangular depression for the bigger, flatter lanyard loop

Total production of Walther/Zella-Mehlis Zero Series P.38 pistols: about 10,200 pistols

P.38, Code "480" 1st Variation (8/1940 – 9/1940)

SN 1 - 6830 (about 6800 pistols)

Characteristics: as before, but

- Slide legend on left hand side: no more Walther banner, but numerical code "480" (for: Waffenfabrik Carl Walther Zella-Mehlis) - horizontally on left hand side of slide (next to "P.38") and on trigger guard
- Grip plates: most pistols with the horizontally grooved, mottled brown to black plastic with a rectangular depression for the lanyard loop
- Lanyard loop: still the smaller, more rounded type.
- Magazines with SN on left hand side of body and two acceptance stamps "Eagle over 359" on spine, spare magazines with an additional stamp "+" next to SN
- Two examples known with slide legend "480 40" ("40" stands for the year of production): SN 3419 and 3478; remarks: Some early P.38 carried the Walther banner in addition to the code designation "480". As this way the code could have been decrypted, the code "480" was milled out leaving a small oval shaped depression.

P.38, Code "480" 2nd Variation (9/1940 – 10/1940)

SN 6831 - 7383 (about 410 pistols)

Characteristics: as before, but

- Most pistols with the bigger, flatter lanyard loop, but the old more rounded and smaller loop was still used, too.

P.38, Code "ac" no date (10/1940 – 11/1940)

SN 7384 - 9950 (about 2560 pistols)

Characteristics: as before, but

- No more Walther banner, but code designation "ac" (for: Waffenfabrik Carl Walther Zella-Mehlis) - horizontally on left hand side of slide (next to "P.38") and on trigger guard
- Magazines with "P.38" and SN on left hand side of body, two acceptance stamps "Eagle over 359" on spine

P.38, Code "ac 40" date hand-stamped (11/1940)

SN 9951 - 5942a (about 6000 pistols)

Characteristics: as before, but

- Year designation "40" added by hand-stamp (below the code stamp "ac" – i.e. asymmetrically) – see illustration 7

Remark:

Up to October 1943 Walther P.38s have the slide legend in two lines (see illustration 7). Due to technical printing reasons these legends are written in one single line in this text.

P.38, Code "ac 40" (12/1940)

SN 5943a - 9965b (about 14 000 pistols)

Characteristics: as before, but

- Year designation "40" no longer added by hand stamp (below the code stamp "ac" – i.e. asymmetrically), but symmetrically on the same height as "P.38", because the date "40" is already integrated in the roll-stamp – see illustration 7

Picture 49: *Walther P.38, code ac 41, 3rd variation, SN 4181j, caliber 9 mm Luger, produced by Walther/Zella-Mehlis in December 1941, horizontally grooved grip plates with rectangular depression for lanyard loop*

P.38, Code "ac 41" 1st Variation (1/1941 – 3/1941)

SN 1 - 4833b (about 24 800 pistols)

Characteristics: as before, but

- Slide serrations: now 15 instead of 14 on either side of slide (standard for P.38 pistols from now on)
- Magazines marked "P.38" on left hand side of body, SN across on floor plate, only one acceptance stamp "Eagle over 359" on spine

P.38, Code "ac 41" 2nd Variation (3/1941 – 11/1941)

SN 4834b - 4530i (about 70 000 pistols)

Characteristics: as before, but

- Code "ac" no longer stamped on trigger guard, but solely on left hand side of slide
- Some examples are known in this SN range with the "death's head" stamp on right hand side of slide – "Waffen SS"?
- Some examples carry the "Reichsfinanzverwaltung" stamp ("R.F.V." plus inventory number), although the standard side arm of the R.F.V was the Walther PP/PPK – see there.

P.38, Code "ac 41" 3rd Variation (11/1941 – 12/1941)

SN 4531i - 9580j (about 15 000 pistols)

Characteristics: as before, but

- Most pistols with the dull "military" finish, as the polishing before the blueing process was omitted (from the end of 1941 by order of the "Reichsminister für Bewaffnung und Munition") – cf. Walther PP/PPK

Remark:

From 1941 onwards – with only a few

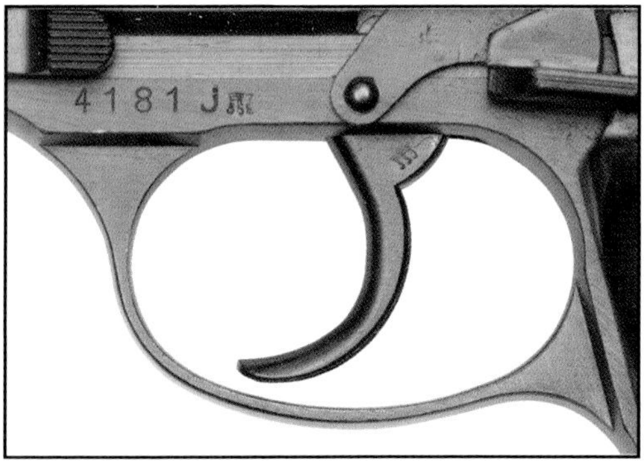

Picture 50: *Walther P.38 frame unreinforced, i.e. frame is flat just below the trigger axle (old type)*

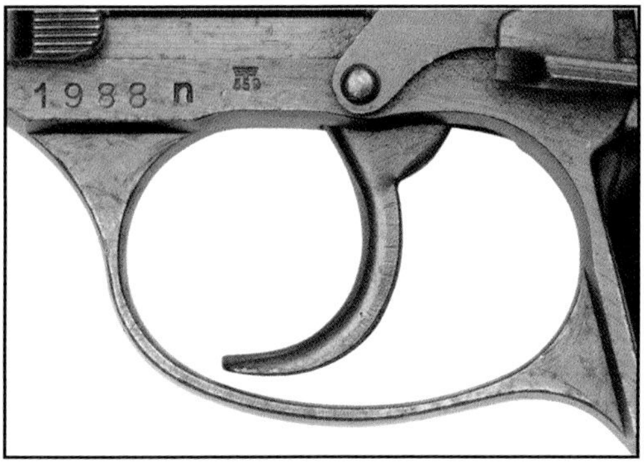

Picture 51: *Walther P.38 frame reinforced, i.e. frame has a hump just below the trigger axle (new type)*

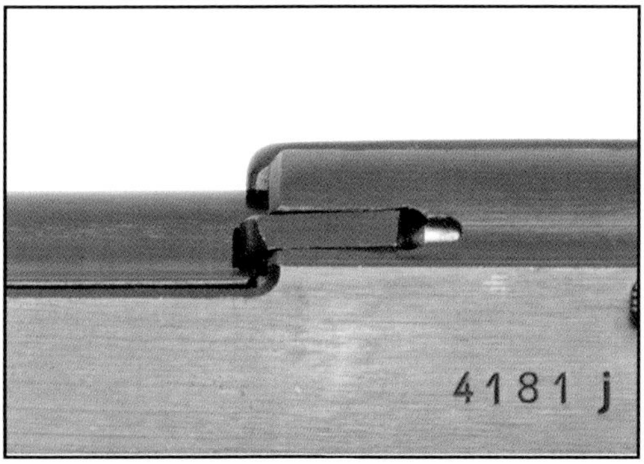

Picture 52: *Walther P.38 slide with a secondary cutout behind the extractor, i.e. a long relief slot (old type)*

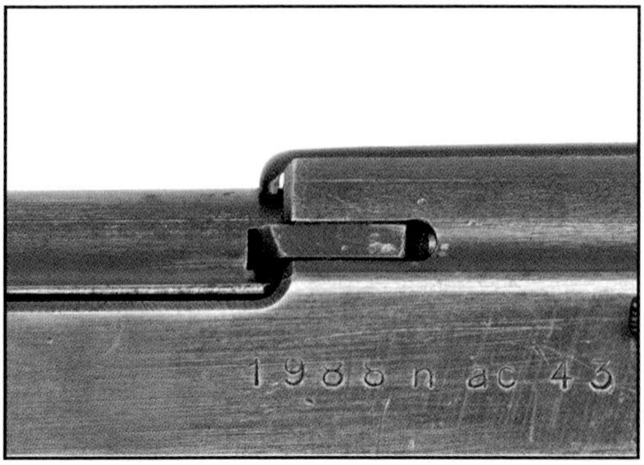

Picture 53: *Walther P.38 slide without a secondary cutout behind the extractor, i.e. a short relief slot (new type)*

9880h produced by the Belgian company "FN" in Herstal (code stamp: "Mi", "mI" or "m"; acceptance stamp "Eagle over WaA 140")

- Some frames produced by the Belgian company "FN" in Herstal (code stamp "Mi", "mI" or "m" on left hand side of trigger guard; acceptance stamp "Eagle over WaA 140")
- Slide hold-open lever mostly no longer grooved
- Finish mostly rougher and sometimes reddish-matt
- Manufacturing tolerances mostly less strict
- Magazines from September 1944 mostly marked with the letter "u" (for "ungehärtet", in English: not hardened; "hardened" refers to the magazine lips only!)

P.38, Code "ac 43" or "ac 44"

Characteristics:

- Code designation on slide written in two lines
- Model designation on slide "P38" (without the dot!)

Remarks:

These slides with "double-line" code inscription were manufactured by FN in Herstal during the German occupation. So they carry the acceptance stamp "Eagle over WaA 140" and one of the code stamps for FN ("Mi", "mI" or "m").

FN produced frames as well with the same acceptance and code stamps (the latter on left hand side of trigger guard).

These frames were shipped to Walther or Mauser and used in the series production of P.38 pistols there. Walther used mostly FN frames, whereas Mauser used mostly FN slides.

P.38, Code "ac 45" (1/1945 – 3/1945)

SN 1 - 2800c (about 32 000 pistols)

Characteristics: as before, but

- Slides with "one-line" code inscription
- From a-block: markings mostly incomplete
- In a-block: some pistols with SN suffixed with capital letter "A"
- Some pistols hammer spur with three coarse grooves on top (instead of the standard eight grooves) – cf. Spreewerk P.38s
- Some frames produced by the Belgian company "FN" in Herstal (code stamp "Mi", "mI" or "m" on left hand side of trigger guard, acceptance stamp "Eagle over WaA 140")
- Some barrels produced by the Czech company "Böhmische Waffenwerke" (Bohemian weapons factory) in Strakonice (code stamp "fnh", acceptance stamp "Eagle over WaA 76")
- Some grip plates made of "Durofol" (see above) or made of polyamide plastic (cf. Mauser P.38s) – maybe the latter were interchanged later.

P.38, Code "ac 45", Second Zero Series (3/1945 – 4/1945)

SN 025 815 - 027 800 (about 2000 pistols)

Characteristics: as before, but

- SN beginning with "0" – cf. first Zero Series in 1939/40

P.38, Code "ac" (3/1945 – 4/1945)

SN unknown (about 3000 pistols)

Characteristics:

- Code "ac" on frame only
- Assembled from Walther frames and parts from the other two P.38 manufacturers

Remarks:

On 9th April 1945 the Walther factory was occupied by General Patton's 3rd Army. After that some P.38 pistols were assembled by Walther workers for US soldiers from parts on stock – mostly with differing serial numbers or no serial numbers at all (so-called "GI-assemblies" or "cigarette guns").

Other Variations and Peculiarities

- Engraved P.38 pistols are known, but were presumably engraved either before April 1945 by Thuringian engravers as Walther subcontractors or engraved after April 1945 by other engravers – cf. engraved PP/PPK pistols above.
- P.38 pistols with acceptance stamps of the German navy ("Eagle/Swastika/M" or "Anchor/M") are presumably not authentic.
- P.38 pistols with shortened barrels or grips are either experimental pistols or they were subsequently shortened in the workshops of the East German "Ministerium für Staatssicherheit" ("Ministry of State Security" of the former GDR, short: "Stasi"). There ten pre-1945 made P.38s were modified at the request of "Stasi" general Kurt Voigt. The frames, the grip plates, the magazines and the barrels were shortened. A wooden magazine extension was added - see chapter "P.38 GDR" and picture 58.
- P.38 pistols with lightweight alloy frames are experimental pistols only.
- One lightweight P.38 (SN 3090) was shipped on 20th October 1941 to the German "Luftwaffe" (air force), but there was no series production of lightweight P.38s. Only the later Walther/Ulm P38s resp. P1s have a lightweight frame.
- Silencers for the P.38s were on the one hand listed in "Geräteliste, Teil 2" (device list, part 2) of the German "Wehrmacht" dated 15th July 1943 as accessory number 1-2019--601, but on the other hand no example is known so far. There was, however, a type of 9 mm Luger ammunition for silenced pistols with the military designation "Pistolennahpatrone 08".

Total production of Walther/Zella-Mehlis P.38 pistols: about 575,200 pistols

Walther P.38 Mauser/Oberndorf Production

According to a government directive early in 1940 all weapon factories in Germany and the occupied territories should be coordinated by a commission and cooperate more closely in future. This should be done to meet the growing demands of the "Wehrmacht" (armed forces). Next to Walther/ Zella-Mehlis the company Mauser/Oberndorf should manufacture P.38 pistols. Mauser was, however, able to do so fully only after the production of Luger P08 pistols had ended in October 1940.

Walther provided the technical drawings, specifications and the "know-how" to be adapted to the Mauser machinery and production processes.

Walther provided sample pistols and some production machinery as well. Extensive development work and test series by Mauser were thus largely unnecessary. So there was no pre-production Zero Series either. Improvements in design and technical changes in the running Walther P.38 production were passed on to Mauser and incorporated in their production almost immediately.

All parts of the Mauser P.38 were mainly made by Mauser in their Oberndorf plant. Only a few parts were provided by subcontractors. Today one would perhaps speak of an in-house production with a high vertical integration. This was done with all the advantages and disadvantages of such a production scheme concentrating on one single location.

Mauser P.38 Grip Plates

The plastic grip plates for Mauser P.38 pistols were provided by Walther or AEG/ Henningsdorf (markings see grip plates for Walther P.38 pistols above).

From 1944 Mauser received glossy, black grip plates from a so far unknown supplier, which were made of polyamide-6 plastic. These have no markings inside.

Mauser P.38 Magazines

The majority of magazines for the Mauser P.38 pistols was produced by Mauser up to the end of the World War II. In addition from 1943 onwards magazines were provided increasingly by the Czech Company "Erste Nordbömische Metallwarenfabrik" (First North Bohemian metal wares factory) in Niedereinsiedel/Sudetenland (code stamp "jvd", acceptance stamp "Eagle over WaA 706").

07/1940 – 6/1942: Test production to set the machines and train the staff.

In 1941 Mauser was given the code designation "byf".

08/1942: Start of pre- production first with slides, barrels etc., which were put into stores without serial numbers. There was a relatively high reject rate with this production run, however.

10/1942: Delivery of the first 700 Mauser P.38 pistols to the "Wehrmacht" after acceptance by the "Waffenamt" (army ordnance office), stamping "Eagle over 135" on pistols, which had passed.

The Mauser code designation for 1942 was "byf 42" – see illustration 9. Some pistols perhaps had "byf 43" already.

01/1943: Start of full series production with code "byf 43".

12/1944: Mauser was given the new code designation "svw" resp. "SVW".

Proof and Acceptance of Mauser made P.38 Pistols by "Heereswaffenamt" ("WaA 135")

Basically the proof and acceptance procedure was the same as the one for Walther made P.38 pistols – see there. The "Waffenamt" acceptance stamp for Mauser made P.38 pistols was "Eagle over 135". The military proof stamp was "Eagle over Swastika".

Acceptance and proof stamps were applied in the same manner and in the same locations as on Walther made P.38s – see there.

As a rule the code designations "byf" resp. "svw" and the year code (last two digits of year) were applied in two lines on the left hand side of the slide (between the model designation "P.38" and the serial number) – see illustration 9.

Mauser In-house-inspection and Maintenance Marks

As on Walther made P.38 pistols one can find numbers, letters or geometric figures on Mauser made P.38s. Some late Mauser P.38s with svw code have a "cloverleaf" mark under the grip plate. The exact meaning of most of these stamps is still unknown. Some of them are probably only repair or maintenance marks.

One stamp could be identified though: "WR" for "Werksrevision" (works revision) – a Mauser in-house inspection stamp – see illustration 12.

Mauser P.38 Serial Number System

Mauser used a system, which was similar to the one Walther used:

- The numbering began with "1" up to "9999", then went back to "1" suffixed with a letter (beginning with "a" through the alphabet) – i.e. "1a" to "9999a", "1b" to "9999b", etc.
- The blocks "1 – 9999" were, however, not necessarily maintained in a strict sequential order. At the turn of the year or after changes in production the numbering did not carry on with a new "1" and the next letter suffix, however – cf. Walther made P.38s.
- The serial numbers were stamped on the left hand side of the slide slightly higher than on Walther made P.38s (right in front of the safety lever, when in the position "fire").

P.38, Code "byf 42" resp. "byf 43" 1st Variation (10/1942 – 4/1943)

SN 1 - 3000f (about 63 000 pistols)

Preliminary remarks:

Mauser made P.38 pistols with the code designation "byf 42" were mostly assembled as late as 1943 using parts manufactured in the previous year – see above. Thus slides marked "byf 42" were used up in 1943 when manufacturing the first 20 000 Mauser P.38s. As it is not possible to draw a demarcation line, both variations are subsumed here as one single variation.

Characteristics: as 1st variation Walther made P.38 "ac 42" – see there, but

- Dull "military" finish throughout
- Suffix letters first in small Gothic letters, later in Latin cursive letters
- Magazines and grip plates not serial-numbered
- From SN 1 to about 1000: most minor parts still stamped ""Eagle over 135"; from about SN 1000: minor parts no longer stamped with this acceptance stamp
- Magazines: first pistols with model designation "P.38", but without the SN on left hand side of body and two "Waffenamt" acceptance stamps "Eagle

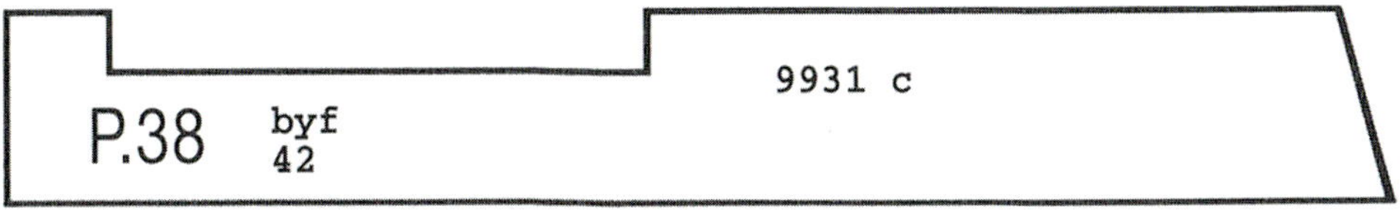

Illustration 9: *example of a Mauser P.38 slide legend with code "byf 42"*

Picture 54: *Walther P.38, code byf 42, 1st variation, SN 2697b, caliber 9 mm Luger, produced by Mauser/Oberndorf in January 1943, horizontally grooved plastic grip plates with rectangular depression for lanyard loop*

over 135" on spine (mostly upside-down, standard from now on); later pistols with one acceptance stamp "Eagle over WaA135" and the legend "P38v" on left hand side of body; magazines with the designation "P.38v" are about 2 mm longer, but this has no effect on the function of the magazines. It may be assumed that "v" stands for "verlängert" (in English: lengthened) or "verändert" (in English: modified) – cf. Walther P.38 magazines.

- Plastic grip plates in dark brown, greenish brown or even black

P.38, Code "byf 43" 2nd Variation (4/1943 –11/1943)

SN 3001f - 1150q (about 108 150 pistols)

Characteristics: as before, but

- Frame reinforced, i.e. frame has a hump just below the trigger axle; standard from now on (cf. Walther made P.38s from July 1943) – see Pictures 50 and 51
- Slide with a modified milling: no longer a secondary cutout behind the extractor, i.e. a short relief slot; standard from now on (cf. Walther made P.38s from July 1943) – see Pictures 52 and 53
- Some pistols with magazines produced by

the Czech company "Erste Nordböhmische Metallwarenfabrik" (First North Bohemian metal wares factory) in Niedereinsiedel/ Sudetenland (code stamp "jvd", acceptance stamp "Eagle over WaA 706")

P.38, Code "byf 43" 3rd Variation (12/1943)

SN 1151q - 8500q (about 5 300 pistols)

Characteristics: as before, but

- Pistols mostly assembled from older parts mixed with later ones with different finishes and various markings
- Slides first stamped on right hand side: "Eagle over 135", "Eagle over Swastika" and again "Eagle over 135", later with: "Eagle over 135", "Eagle over Swastika" and "Eagle over WaA 135"

P.38, Code "byf 44" 1st Variation (1/1944 – 10/1944)

SN 8501q - 5000c (about 116 000 pistols)

Characteristics: as before, but

- Serial number continue consecutively from the previous year
- "Waffenamt" acceptance stamp increasingly "Eagle over WaA 135" (instead of "Eagle over 135")
- Rougher surfaces due to less careful workmanship (e.g. traces of lathes on barrels etc.)
- Metal finish: often so-called "dual tone" – i.e. dull blue and grey phosphated parts mixed

P.38, Code "byf 44" 2nd Variation (11/1944 – 12/1944)

SN 5001c - 5000d (about 10 000 pistols)

Characteristics: as before, but

- New phosphated finish on most or even all parts (Due to this greyish color these P.38s are often called "grey ghosts" in the US literature)
- Some pistols in the "dual tone" finish, if older parts or parts provided by subcontractors (e.g. barrels) were used in production
- Slide hold-open lever no longer grooved
- Magazines increasingly with the stamp "u" (below the model designation), "u" for "ungehärtet" – i.e. magazine lips were no longer hardened

Remarks:

From 6th December 1944 onwards the code designation for the Mauser production was changed from "byf" to "svw" resp. "SVW". Before the end of 1944 there were, however, only a few pistols made still with that new code.

In 1944/45 some clever Mauser engineers even designed a special device for test firing P.38 pistols: "Schießmaschine für Pistolen" (project numbers 44 98 43, 44 98 52 resp. 44 98 55) with its own optical aiming device (project number 501.0-52V)

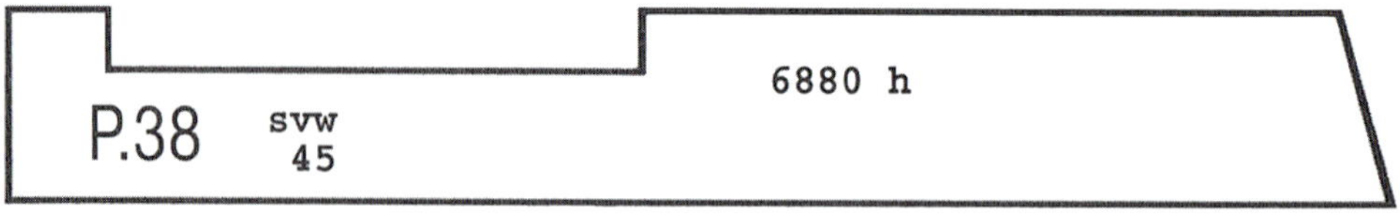

Illustration 10: *example of a Mauser P.38 slide legend with code "svw"*

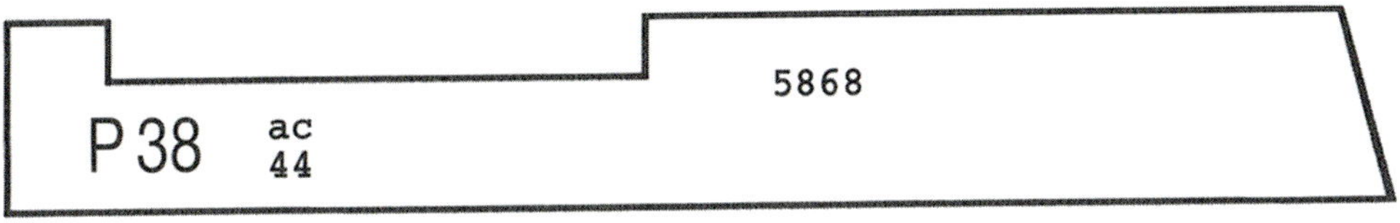

Illustration 11: *example of a Mauser P.38 slide legend for the German police with code "ac 44" (FN made slide)*

P.38, Code "svw 45" German Variation (1/1945 – 4/1945)

SN 5001d - 2200f (about 17 000 pistols)

Characteristics: as before, but

- Metal finish: predominantly grey phosphated finish, sometimes mixed with blued parts (Due to this greyish color these P.38s are often called "grey ghosts" in the US literature, too)
- Some pistols with FN made slides (code stamp "Mi", "mI" or "m", acceptance stamp "Eagle over WaA 140"), rarely with FN made frames (code stamp "Mi", "mI" or "m" on left hand side of trigger guard, acceptance stamp "Eagle over WaA 140")
- Grip plates made by AEG or made of glossy black polyamide-6 plastic (see above). There is no proof that phosphated sheet metal grip plates were used. Such grip plates were allegedly designed and produced by Mauser already, but obviously not yet used in assembling Mauser P.38 pistols.
- Towards the end of World War Two more and more parts with mismatched SNs or without markings resp. acceptance stamps were used in production.

Mauser P.38 Pistols for the German Police

Walther and Spreewerk produced P.38 pistols exclusively for the German "Wehrmacht" and their allies. Mauser manufactured P.38 pistols for the German military and the police as well. The latter were issued to uniformed police units, which up to then had carried other side arms in caliber 9 mm Luger – e.g. the Luger pistol P08.

P.38, Code "byf 43" to "svw 45" (end of 1945 – beginning of 1945)

code "byf 43":	about SN 1 - 1400
code "byf 44":	about SN 1400 - 8780
code "svw 45":	about SN 5611 - 5900
	(all together about 17 000 pistols)

Characteristics: as P.38 for the "Wehrmacht" of the comparable production period, but

- Some pistols with slide legend "ac 43" or "ac 44", made by FN and not by Walther though; model designation "P38" (without a dot!); code designations written in two lines
- Commercial proof marks (on left hand side of slide): "Eagle over N"
- From about SN 1000: one "Waffenamt" acceptance stamp on right hand side of slide (The acceptance of police pistols was carried out by "Waffenamt" officers in assistance, but under less strict rules)
- Police acceptance stamps:
- From "byf 43" SN 1 to "byf 44" SN 4308: "Eagle over Cross or Swastika"
- With suffix letter "L",
- From "byf 44" SN 4309 to "svw 45" SN 5900: "Eagle over Cross or Swastika"
- With suffix letter "F"
- Metal finish: as P.38s for the "Wehrmacht"

of the comparable production period, i.e. blue-black blued, "dual tone" or grey phosphated finish

Total production of Mauser/Oberndorf P.38 pistols (up to 20th April 1945): about 326,350 pistols

Remarks:

On 22nd February 1945 the Mauser plant was bombed and the major part of the energy supply facilities was destroyed. On 20th April 1945 the Mauser plant was occupied by French troops. The production facilities were not destroyed or closed, however, but production carried on under French administration. From then onwards Mauser produced P.38 pistols for the French military and police forces only.

First P.38 parts out of storage were used – some of these with Mauser, Walther, Spreewerk or no markings at all. Parts from captured or cannibalized P.38 pistols were used as well.

The serial number system was retained.

P.38, Code "svw 45" French Variation (4/1945 – 12/1945)

SN blocks "f" to "k" (about 40 000 pistols)

Characteristics: as before, but

- Parts from old stocks were used in production. The old serial number system was retained. So sometimes existing serial numbers were either overstamped or removed by grinding or milling.
- Suffix letters to SN sometimes capital letters
- P.38s for the French military: French military acceptance stamp ("five-pointed star" – often named "French star") next to the German acceptance marks on right hand side of slide
- P.38s for the French police: : French police acceptance stamp ("bomb with three flames") next to the German acceptance marks
- Stamp "three leaf clover": meaning unknown; such a symbol is often associated with the Celtic culture ("shamrock") though
- Metal finish: either
  - a) grey "German" phosphated finish or
  - b) "dual tone" (phosphated and blued parts mixed) or
  - c) dark grey "French" phosphated finish (at least on newly manufactured parts) or
  - d) matte-black blueing on all parts (mostly on P.38s for the French police) or
  - e) polished dark blue finish (mostly on P.38s for the French police, too) or
  - f) old finish refurbished with resulting discoloring (on sheet metal grip plates, too)
- Grip plates: mainly the typical sheet metal, some remaining polyamide-6-grips
- Magazines mostly from old German stock (stamped "P.38v", the additional letter "U" and only one "Waffenamt" acceptance stamp – see above)

P.38, Code "svw 46" French Variation (1/1946 – 6/1946)

SN blocks "k" to "l" (about 8 500 pistols)

Characteristics: as before, but

- No more technical or other design changes
- SN range 11 – 5001 : no military French acceptance stamp ("five-pointed star"), as these P.38 pistols were assembled as late as by the beginning of 1947 by "Manufacture d'Armes Chachtelleraut" (MAC) from parts, which were provided by Mauser/Oberndorf. These parts were assembled to complete pistols there and then issued to the Ministry of the Interior for the national police.

Picture 55: *Walther P.38, code svw 46, 1st variation, SN 7641k, caliber 9 mm Luger, produced by Mauser/Oberndorf in January 1946, "French" variation with phosphated finish, "French star" (on right hand side of slide), horizontally grooved metal grip plates (rectangular depression for lanyard loop), later post-1945 use by Austrian army ("Bundesheer", stamp "BH")*

Illustration 12: *stamp "WR" for Mauser "Werksrevision"*

Other Variations and Peculiarities

- Mauser made P.38 pistols with the Mauser banner (so-called "Mauser barrel") are most likely not authentic.
- Some pistols carry the stamp "WR" for "Werksrevision" (works revision) – a Mauser in-house inspection stamp – see illustration 12.

Remarks:

In May 1946 the Mauser/Oberndorf plant was at least partially under French control. In those days the above mentioned P.38 pistols (serial number range block "k" and "l") were assembled from parts from old stocks. These pistols did not comply with the usual security standards.

All in all about 37 854 P.38 pistols were made under French administration from 1945 to 1946. If one takes into account the serial number blocks "f" to "l", this should actually be a total of 65 000 pistols. That discrepancy may be explained by the high reject rate though.

After massive Soviet protests the French had to stop their P.38 pistol production in the Mauser/Oberndorf plant, as this was in contradiction to the "Potsdam agreement" of the four victorious allies in August 1945.

According to the Allied Council directive number 16 from 6th November 1945 the West Berlin police was not allowed to carry side arms made in Germany. So they were first issued "French" Mauser made P.38 pistols (code "svw 45" and "svw 46") from 1946 onwards. In 1966 these police forces then procured "French" P.38 pistols with Manurhin inscriptions.

Walther P.38 Spreewerk/Grottau Production

The company named Metallwarenfabrik Spreewerk Berlin-Spandau was separated from the "Deutsche Industriewerke ("DIW", in English: "German industrial plants") in 1935. Although the company address was Berlin-Spandau, the pistol production was carried out in the branch factory in Grottau on the Neisse River (in Czech: Hradek nad Nisou) in so-called "Sudetenland". A producer of grip plates, the company Julius Posselt ("Wehrmacht" code designation "1W"), was situated in nearby Gablonz. Other subcontractors for minor parts were the Czech companies "Böhmische Waffenwerke" (Bohemian weapons factory) in Strakonice (code stamp "fnh", acceptance stamp "Eagle over WaA 76") and "Erste Nordböhmische Metallwarenfabrik" (First North Bohemian metal wares factory) in Niedereinsiedel/ Sudetenland (code stamp "jvd", acceptance stamp "Eagle over WaA 706"). Today one would perhaps speak of an industrial cluster in "Sudentenland".

Next to P.38 pistols Spreewerk produced the "Volksgewehr" ("people's rifle") model "VG2" in Grottau, too.

Remarks:

The name of the Spreewerk company is often cited wrongly – especially in the US literature – as "Spreewerke" in the plural form. The singular form without an "e" is correct though.

Walther/Zella-Mehlis organised, controlled and, if necessary, corrected the work of Spreewerk/Grottau. From May to June 1942 three employees were sent from Grottau to Zella-Mehlis to be trained there. The first 50 Spreewerk P.38 pistols were presumably assembled then. In July/August 1942 the series production of Spreewerk P.38 pistols started in Grottau.

As was the case with Mauser, too, improvements in design and technical changes in the running Walther P.38 production were passed on to Spreewerk and incorporated in their production almost immediately. Spreewerk/Grottau produced P.38 pistols for the "Wehrmacht" only.

Compared to the P.38 pistols made by Walther or Mauser the ones made by Spreewerk are mostly of inferior quality. Some specimens show an excellent workmanship though, proving that Spreewerk could do better as well. Even within the same serial number block the quality may differ immensely.

The design characteristics of Spreewerk P.38 pistols are totally identical to the ones made by Walther or Mauser.

The "Wehrmacht" expected a monthly delivery of 10 000 Spreewerk P.38 pistols. This target could not be met by Spreewerk, however. This demand could be roughly fulfilled only from April 1943 to April 1945. It has to be taken into account, however, that not all the P.38 pistols produced by Spreewerk could be shipped to the "Wehrmacht". The reject rate by the acceptance officers was especially high with the Spreewerk production.

A very detailed and outstanding survey of the Spreewerk P.38 production was published in Jan Balcar, "Die Spreewerk P.38 – die Geschichte des Grottauer Werkes und seiner Kriegsproduktion", Pilsen 2009. The Czech author could do his research on-site and even interview living witnesses. The following statements as to production data and serial numbers are largely based on his research.

Proof and Acceptance of Spreewerk made P.38 Pistols by "Heereswaffenamt" ("WaA 88")

Basically the proof and acceptance procedure was the same as the one for Walther or Mauser made P.38 pistols – see there. The "Waffenamt" acceptance stamp for Spreewerk made P.38 pistols was "Eagle over 88". The military proof stamp was "Eagle over Swastika".

Acceptance and proof stamps were applied in the same manner and in the same locations as on Walther made P.38s – see there.

Unlike with Walther and Mauser made P.38 pistols there is no year code (last two digits of year) on Spreewerk P.38 slides – see illustration 13.

The "Waffenamt" officer in charge of accepting Spreewerk P.38 pistols was Max Schneider with up to seven fellow officers.

The code designation for Spreewerk P.38 pistols was "cyq". This code stamp was applied on the left hand side of the slide between the model designation and the serial number.

As late Spreewerk P.38s carry the code designation "cvq", there was a heated discussion in the Walther literature: The company "Waggonfabrik Josef Rathgeber" in Munich was actually given this code designation, but Rathgeber produced mainly wagons for the German railways. So it was discussed whether the Spreewerk P.38 production was moved to Munich because of the approaching Soviet troops. Another explanation could have been that Spreewerk was given a new code around 1944 – cf. the change of the Mauser codes from "byf" to "svw".

This dispute could be settled by Jan Balcar: All Spreewerk P.38 parts marked "cvq" were made in Grottau. There were several dies used when applying the acceptance stamps on Spreewerke P.38 pistols. In one these dies used by one the acceptance officers the lower part of the letter "y" had broken off due to a faulty hardening and thus this letter could be read as a "v".

Three-digit numbers under the left hand side grip plate are Spreewerk lot or charge numbers.

Spreewerk P.38 Grip Plates

The grip plates for the first run of 20 000 Spreewerk P.38 pistols were provided by Walther/Zella-Mehlis. They were actually made by AEG/Henningsdorf for Walther P.38 pistols, so some of them carry the Walther "Waffenamt" acceptance stamp "Eagle/359".

All later grip plates for Spreewerk P.38 pistols were made by the company Julius Posselt in Gablonz.

Outside these show the following peculiarity: out of 31 horizontal grooves only five are interrupted near the grip screw (cf. Walther grip plates).

Inside these grip plates are marked with three circles containing (from top to bottom): the MPBD logo with "1 W" (code for Posselt) and "31" or "41" (code numbers for the material and additives); the middle circle is empty; numbers designating the position in the casting mold.

The Posselt grip plates show colors varying from brown to reddish brown or black brown.

Spreewerk P.38 Magazines

The magazines for Spreewerk P.38 pistols were provided either by the Czech company "Erste Nordböhmische Metallwarenfabrik" (First North Bohemian metal wares factory) in Niedereinsiedel/Sudetenland (code stamp "jvd", acceptance stamp "Eagle over WaA 706") or by a so far unknown provider without a code stamp. This could have been the Spreewerk subsidiary in the neighboring town of Zittau or another subsidiary in Kratzau. This theory still needs evidence though.

Spreewerk In-house-inspection and Maintenance Marks

As on Walther made P.38 pistols one can find numbers, letters or geometric figures on Spreewerk made P.38s. The exact meaning of most of these stamps is still unknown. Some of them are probably only repair or maintenance marks.

An exception here is the so-called "asterisk" (a little star with a hole in the middle, which is similar to the Germanic "hagal rune") – cf. Walther made P.38 pistols:

One theory states that this stamp denotes parts which were found not to be dimensionally accurate or badly finished, but repairable.

Recent research – especially by Jan Balcar from the Czech Republic – has shown that this theory is right: Parts marked with the "asterisk" were rejected, refurbished and then used in series production.

Spreewerk P.38 Serial Number System

The Walther serial number system (1 – 9999, 1a – 9999a, 1b – 9999b, ...) was adopted by Spreewerk, too. In the first production run the serial numbers were suffixed with a letter, as with Walther or Mauser made P.38s. After SN 9999z (February 1945), however, the serial numbers were prefixed with a letter starting with "a" again. After SN b5000 the serial numbers were prefixed with the number "0" ("Zero Series"). The very last Spreewerk P.38s had serial numbers, which were prefixed with "00" ("Double Zero Series").

The letter "q" was obviously not used as a suffix or prefix.

A pre-production "Zero Series" was obviously not necessary, as the test runs had been made by Walther already. There is one pistol known with SN 02, however.

P.38, Code "cyq" 1st Variation (7/1941 – 12/1942)

SN 1 - about 7000 (about 7000 pistols)

Characteristics: as Walther or Mauser made P.38 of the same period – see there, but

- SN 1 to 50: The very first Spreewerk P.38 pistols, which were made by three Spreewerk employees while being trained in the Walther factory, were obviously not accepted by "Waffenamt" officers – see above. Their whereabouts are unknown. Maybe they were destroyed. The serial numbers might have been prefixed with a "0" ("Zero Series") – cf. Spreewerk P.38 with SN 02 (see above)
- Some very early pistols: code stamp "cyq" applied vertically
- Up to SN 5000: parts made by Walther/ Zella-Mehlis still used when assembling Spreewerk P.38 pistols, some magazines without a code designation (made by other Spreewerk subsidiaries in Zittau or Kratzau?) and with one "Waffenamt"

Picture 56: *Walther P.38, code cyq, 3rd variation, SN 2940u, caliber 9 mm Luger, produced by Spreewerk/Grottau in 1944, hammer spur with three groves, horizontally grooved Posselt made grip plates with rectangular depression for lanyard loop*

acceptance mark "Eagle over 88" only or magazines made by "Erste Nordböhmische Metallwarenfabrik" (First North Bohemian metal wares factory) in Niedereinsiedel/ Sudetenland (code stamp "jvd", acceptance stamp "Eagle over WaA 706")

- Metal finish gradually changes from a polished blue to a dull "military" finish

P.38, Code "cyq" 2nd Variation (1/1943 – 7/1943)

SN 7001 - about 9100d (about 42 100 pistols)

Characteristics: as before, but

- Workmanship and quality increasingly worse; tolerances and fits worse, too
- From SN 1000d: magazines with "P.38" on left hand side of body, only one "Waffenamt" acceptance stamp "Eagle over 88" on spine
- From SN 1584d: frame reinforced, i.e. frame has a hump just below the trigger axle; standard from now on (cf. Walther made P.38s from July 1943) – see Pictures 50 and 51
- From SN 1584d: slide with a modified milling: no longer a secondary cutout behind the extractor, i.e. a short relief slot; standard from now on (cf. Walther made P.38s from July 1943) – see Pictures 52 and 53
- From about SN 1600d: magazines with "P.38" and "jvd" on left hand side of body, only one "Waffenamt" acceptance stamp "Eagle over 88" on spine
- Some stamps not precisely applied, i.e. by angled application etc.
- Grip plates: brown, black brown or other colors possible

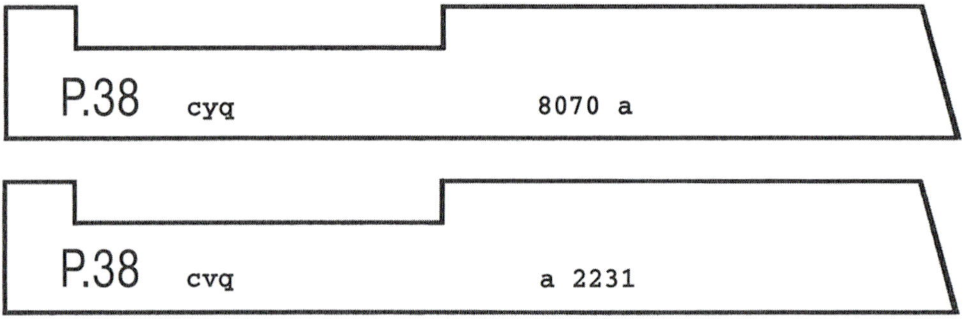

Illustration 13: *Spreewerk P.38 slide legends with suffixed SN (above) and prefixed SN (below)*

P.38, Code "cyq" 3rd Variation (7/1943 – 2/1945)

SN 9101d - about 5000z (about 190 250 pistols)

Characteristics: as before, but

- Workmanship and quality increasingly worse, tolerances and fits, too
- From SN block "j" some pistols and from SN block "z" most pistols with a hammer spur with three grooves on top (instead of the standard eight grooves)
- From SN block "l": slide hold-open lever without grooves
- From SN block "v": magazines increasingly marked with "u" for "ungehärtet" – i.e. magazine lips not hardened
- In SN block "x": about 1400 pistols with FN made slides (acceptance stamp "Eagle over 140", code stamp "Mi", "mI" or "m")

P.38, Code "cvq" 1st Variation (2/1945)

SN about 5000z - about 10 000z (about 5000 pistols)

Characteristics: as before, but

- Code stamp "cyq" mostly replaced by "cvq", but with a flowing transition (see discussion above for the reasons for this "change")

P.38, Code "cyq" 2nd Variation (2/1945 – 4/1945)

SN a11 - about b5000 (about 15 000 pistols)

Characteristics: as before, but

- SN prefixed by a letter starting with "a" again

P.38, Zero Series (4/1945)

SN 01 - about 06435 (about 6435 pistols?)

Characteristics: as before, but

- Assembled from older parts in stock mixed with newly made ones or parts from cannibalized pistols
- SN prefixed by a "0"
- Some pistols with FN made slides (slide legend "ac 43" or "ac 44"; model designation "P38" – without a dot!; code designations written in two lines, acceptance stamp "Eagle over 140")
- Some pistols without acceptance stamps or acceptance stamps replaced by the capital letter "U" (for "Untersuchung", in English: examination)

- The "Volkssturm" ("people's army") was presumably equipped with these pistols by the NSDAP (Nazi party).

P.38, Double Zero Series (4/1945)

SN 001 - about 00100? (about 100 pistols?)

Characteristics: as before, but

- SN prefixed by a "00"
- Acceptance stamp "Eagle over 88" on barrel only; frame, slide and locking block marked with the capital letter "U"
- Magazines stamped with "P.38 U" on left hand side of body, only one acceptance stamp "Eagle over 88" on spine

Other Variations and Peculiarities

- Spreewerk P.38 pistols with an additional year code are presumably not authentic.
- SN 2169n: experimental pistol with a frame made using precision casting to save costs, additional proof mark "Eagle over Swastika" on left hand side of trigger guard

Remarks:

In May 1945 Soviet troops occupied the Spreewerk plant in Grottau and dismantled the production machines. The latter were unfortunately made unusable due to the inappropriate handling by the Soviet soldiers.

There were, however, about 200 pistols assembled for Soviet soldiers – without acceptance marks though (cf. the "GI-assemblies" in the Walther plant).

In 1946 the Czech government had 3000 P.38 pistols assembled with the code designation "CZ 46" – made from parts on stock (see there).

Total production of Spreewerk/Grottau P.38 pistols: about 260,000 pistols

Walther P.38 GDR (German Democratic Republic) Production

After the allied forces had taken the three Walther P.38 production sites, they – especially the Soviets – were in the possession of large numbers of captured weapons, factory stocks, military and police arsenals and even whole production facilities. Many of these weapons were Walther pistols.

What happened to the Walther/Zella-Mehlis files, documents and production machines is still unknown. On the one hand the US troops were there first, so they had enough time to take away files, but had little time to remove machinery. The Soviets took over later, but maybe most of the files had already been taken away by US troops. Rumor has it that at least the Walther PPK production machines were taken away by the Soviets and were later given to the Communist Chinese – see there.

Due to the Four-Power Agreement there were no armed police or military forces in the Soviet-occupied zone after May 1945. Only the big cities Berlin and Dresden had regular uniformed police ("Ordnungspolizei")

by order of the Soviets in May 1945 already. From mid-1945 onwards police forces were introduced in the newly established federal state and province administrations as regional police authorities.

According to the Allied Council directive number 16 from 6th November 1945 the German police was not allowed to carry side arms made in Germany - with an exception in accordance with clause 6 dealing with "other appropriate weapons". Captured pistols like the Walther PPs, PPKs and P.38s could obviously be subsumed here and thus be used for arming the East German police – cf. pistols for the West Berlin police (see there).

At the end of 1946 the East German police forces were subordinated to the new federal state governments. The "Deutsche Verwaltung des Innern" ("DVdI", roughly translated: administration of the interior) acted in an advisory and coordinating role – e.g. with the setup of the border police. Mid-1948 the police forces were largely decentralized and directly subordinated to the "Deutsche Verwaltung des Innern". The two "Hauptabteilungen" (main departments) "Grenze" (border) and "Bereitschaft" (stand-by or on-call, i.e. riot police) were established.

As late as May 1949 the East German police forces officially named "Volkspolizei" (people's police) were formed. "Volkspolizei" was officially abbreviated "VP" (not "VOPO", as often wrongly but popularly named). At about the same time the "Betriebsschutz" (industrial guards) of the "Volkspolizei" was established.

When the "German Democratic Republic" (GDR) was founded in October 1949, some of the tasks of the "DVdI" were passed on to the "MdI" (Ministry of the Interior). Somewhat later the "Hauptabteilungen" named "Deutsche Volkspolizei" (DVP), "Betriebsschutz" (BS), "Pass- und Meldewesen" ("PM", passport and registration authority), "Ministerium für Staatssicherheit" ("MfS", Ministry of State Security) and the "kasernierte Volkspolizei" ("KVP", roughly translated: people's police in barracks) were established. After the uprising in June 1953 there were groups of "Arbeiterwehr" (workers' guards) founded in large concerns. These were later developed into the "Kampfgruppen der Arbeiterklasse" (roughly translated: fighting groups of the proletariat).

Most of these groups were equipped with former "Wehrmacht" weapons (e.g. Walther P.38 pistols) simply to save costs. As early as mid-1950s, however, the border police and the KVP were already equipped with Soviet weapons, e.g. the Tokarev TT 33 pistol (caliber 7.62 x 25 mm).

In January 1956 the "Kasernierte Volkspolizei" (KVP) became the "Nationale Volksarmee" (NVA, national people's army) with its subdivisions navy, air force and land forces. Gradually these armed forces were equipped with Soviet weapons – e.g. the Makarov pistol (PM) in caliber 9 mm Makarov.

In the 1960s the weapons of the police forces were largely replaced by Soviet models, too. Next to Soviet pistols the following pistol models were newly introduced resp. reused:

- The East German Walther PP copy: "Pistole 1001" in caliber 7.65 mm Browning (see there), which was taken out of service in 1977/78 again
- The Romanian Walther PPK copy: "Md. 74" in caliber 7.65 mm Browning (see there), which was introduced from 1980
- The Czech pistol models CZ Duo and CZ Z (both in caliber 6.35 mm Browning)
- The Walther P.38 pistol

This restructuring of the GDR armament set many P.38 pistols (often called "VOPO P.38s") free. Thus many of them were subsequently sold – to Western states for hard foreign currency. Those sales were mostly handled via "socialist brotherhood" countries – e.g. the Czech Republic and Bulgaria. From there these guns were then shipped to the Western markets. As a compensation transaction the Czech Republic offered the above mentioned CZ pistols for the GDR P.38s.

Characteristics of GDR P.38 Pistols

- These P.38s were often assembled from parts made by different pre-1945 producers – Walther, Mauser or Spreewerk. Hence the markings often were not coherent.
- GDR P.38s were mostly re-blued with a matte finish – including the sear and the locking block, which normally stayed in-the-white. With the early P.38s the high-gloss blueing was only touched up.
- "Wehrmacht" markings, e.g. swastikas, and code designations should actually be defaced by "X"-stamps. This was not done consistently though.

Serial Numbers of GDR P.38 Pistols

- Original serial numbers were retained, if the major parts remained unchanged.
- If the barrel or the slide were replaced, they received the serial number of the frame, i.e. the numbers were mostly overstamped. Sometimes the old serial numbers were grinded or milled out and then the new numbers were stamped in their place.
- If the frame was replaced, the number of the slide was transferred to it and the frame was treated as described above.
- If two magazines were issued with the pistol, they were marked with "1" and "2".
- Newly made P.38 pistols have serial numbers prefixed with the letter "N" (for "Neufertigung"?, in English: new manufacture)

Stamps and Markings on GDR P.38 Pistols (cf. illustration 14, too)

- "Sunburst": This is actually a stylized police or guard star, which encloses a

Illustration 14: *East German markings on P.38 pistols. Inserts from top to bottom: Crown over N, Crown over U, A.B.9, "sunburst", "daisy" (4 mm)*

shield with a number (1 – 30) inside. These numbers designate the "Volkspolizei" district, which the pistol was issued to. The stamps came in two sizes: 2.5 and 5 mm. Sometimes "sunbursts" were stamped without numbers inside. If a pistol was issued to several districts one after the other, several "sunbursts" were applied.

- "Crown over N": proof mark (only on newly manufactured parts), sometimes with a date code – e.g. "753" for July 1953; not to me mixed-up with the pre-1945 "Crown over N" stamp, which shows a slightly different shape
- "Crown over U": inspection stamp on newly manufactured or repaired frames after the final proof firing
- "A.B.9" or sometimes "A.B.70" (both in a wreath): acceptance mark of the "Abnahmebeamte", only on newly manufactured or refurbished parts – cf. "Pistole 1001" (see there)
- "Daisy": This is actually a stylized police or guard star, too, used as a combination stamp on police P.38s after proof firing and a functional test; used as well for overstamping "Wehrmacht" stamps. The name "daisy" was coined by collectors.

Newly manufactured Parts made in Czech Republic for GDR P.38 Pistols

- Barrels: From 1950 onwards manufacture of barrels using in-the-white barrels from old Spreewerk/Grottau stocks; from 1953 onwards new manufacture of barrels for GDR P.38s in "Böhmische Waffenfabrik" in Strakonice (former code stamp "fnh"), shipped to GDR mostly without markings.
- Frames and slides: manufactured from 1949 to 1950 using old Spreewerk/Grottau machines – see Spreewerk/Grottau P.38s
- Magazines: delivered to GDR mostly in-the-white, too; markings stamped in GDR later

Altered Mechanism of GDR P.38 Pistols

- Firing pin: This part was modified with most GDR P.38 pistols – no matter whether the pistol was newly manufactured, or refurbished. The modification consists of two additional lugs towards the end of the firing pin, which should prevent the firing pin from reaching the primer, in case of a fracture of the pin. The slide needed no further modification for this.
- Safety shaft: The lower part of the safety shaft was milled away, so that – in the position "safe" – only the two front lugs meshed with the safety shaft. This brought with it an unnecessary weakening of the shaft and thus a danger for the shooter. And this could lead to a very dangerous situation, if a safety shaft "new type" was accidentally combined with a firing pin "old type": When de-cocking the pistol by rotating the safety lever, the pistol was in the "full automatic fire" mode and fired until the magazine was empty.

New Manufacture of GDR P.38 Pistols

In the second half of 1953 a small series of about 120 pistols was produced with new frames, slides and barrels, but with minor parts from pre-1945 production.

For obvious reasons (Four Powers Agreement, license rights, etc.) this production was kept secret. The manufacture was done in a secret department of the former "Haenel-Werk" in Bahnhofstraße/Suhl. This secret department was called "Spezielle Produktion" ("special production" – i.e. secret military production). At that point in time "Haenel" was already part of the large, state-controlled concern "VEB Ernst-Thälmann-Werke" – cf. "Pistole 1001" production (see there).

The newly manufactured P.38 pistols were accepted by the "Abnahmebeamte" and then stamped "A.B.9" (in a wreath).

Mixed types made of old and newly

Picture 57: *GDR P.38, SN 1000, caliber 9 mm Luger, newly manufactured by "VEB Ernst-Thälmann-Werke" in 1953, no slide legend, no suffix letter "N" yet, horizontally grooved grip plates with rectangular depression for lanyard loop*

Picture 58: *GDR P.38"k", code ac 41, 2nd variation, SN 7884d, caliber 9 mm Luger, produced by Walther/Zella-Mehlis in May 1941, shortened by a "Stasi" workshop in 1953 (barrel, frame, grip plates and magazine shortened, front sight on slide bridge)*

manufactured parts exist as well.

Characteristics of GDR P.38 Pistols (new manufacture): in addition to the modifications described above

- Serial numbers from SN (N)1000 up to about N1120 (hand- or machine-stamped), first examples still without the prefix letter "N"
- Grip plates: mostly made of greenish or black plastic, with horizontal grooves similar to the grip plates from pre-1945 production, no markings inside
- No slide legend on left hand side of slide
- "VOPO-star" (small size)
- Acceptance stamp: "A.B.9" or sometimes "A.B.70" (both in a wreath)

Manufacture of GDR P.38"k" Pistols

At the request of "Stasi" general Kurt Voigt ten pre-1945 P.38 pistols were shortened in the workshops of the East German "Ministerium für Staatssicherheit" ("Ministry of State Security" of the former GDR, short: "Stasi"). The frames, the grip plates, the magazines and the barrels were shortened to make these pistols "P.38k"-type ones. The front sight was installed on the slide bridge (cf. later Walther P38k pistols made by Walther/Ulm). A wooden magazine extension was added - see picture 58.

Major Kurt Voigt was a graduate engineer of ballistics and according to his own confession an avid reader of the West German gun magazine "Deutsches Waffen-Journal".

There is at least one GDR "P.38k" known, which was produced by Spreewerk and was later shortened in one of the "Stasi" workshops with a thread on the muzzle of the barrel for a silencer.

Refurbishing of Mod. HP Pistols in GDR

The Mod. HP pistols procured by the GDR were given the same treatment as the P.38 pistols – see above.

Walther P.38 Post-1945 Czech Republic Production

As in East Germany there were huge amounts of weapons left behind in Czechoslovakia by the surrendering German troops. Next to the captured weapons there were large stocks of P.38 parts on Czech soil – e.g. in Spreewerk/Grottau, Erste Nordböhmische Metallwarenfabrik or Böhmische Waffenfabrik (later named CZ).

If in good condition these captured weapons were proof fired by the Czech company BRNO in Brno and then put in storage there. From 1940 to 1949 the proof mark was a "five-pointed" star. From 1950 to 1953 it was a stylized, X-shaped star with an underscore. Pistols in less good condition were either repaired or cannibalized.

In 1946 P.38 pistols were assembled from existing parts in stock under the direction of the Czech government. This way about 3000 P.38s were "produced" in the former Böhmische Waffenwerke in Strakonice. The parts used in assembly were mostly taken from pre-1945 Walther, Spreewerk, FN or Böhmische Waffenfabrik production.

As most of these P.38 pistols were assembled in 1946, they are often called model "CZ 46" in the Walther literature. This designation appears only on the right hand side of the trigger guard though. With some very early specimens the Czechs bothered to replace the "38" in the slide legend by "46". The code designation on the left hand side of the slide was retained.

As this way enormous amounts of P.38 pistols were available later, these pistols were issued to the Czech armed forces on an experimental basis. The military designation then was "Vz 46" ("Vzor" for model).

Characteristics of Czech made P.38 pistols:

- Markings coherent with the original manufacturer – see above
- Proof mark: Czech "rampant lion" (left facing) – cf. illustration 15
- Acceptance stamp of the Czech procurement office: letter "E"
- Number "46" on right hand side of trigger guard
- If the original serial numbers were still there, they were overstamped by hand with the new ones. The original German serial numbers were very rarely retained.
- Metal finish: dull "military" finish
- Magazines were either hand-stamped or electro-penciled with the serial number; if two magazines were issued with the pistols, they were hand-stamped with "1" or "2"
- Serial number range: 1 – 3000

Czech Exports of P.38 ("CZ 46") and captured P.38 Pistols

Beginning of 1948: Sales of P.38 pistols to Israel were made. These pistols were not modified in any way and all markings were left intact. The P.38 was never officially adopted as a standard service pistol in Israel, however.

1950: Sales of P.38 pistols to the former GDR were made. These pistols were refurbished in the Czech Republic, but stamped in the GDR later. In the second half of the 1950s these P.38 pistols were sold back again to the Czech Republic due to the restructuring of the GDR army and police armament. In the Czech Republic all the GDR markings were defaced. Then these pistols were sold to other countries as war surplus.

It may be assumed that many of the P.38s with the code stamps "ac 44" or "ac 45", which were offered on commercial markets in post-World War Two years, were not produced pre-1945 by German factories. Rumor has it that the Czech government trading company "Omnipol" had these pistols assembled from parts found in old stocks and then sold them on the international market. When the supplies from these old stocks had run out, there were even new parts manufactured on old machines (?). These pistols were then stamped "authentically" with old German dies, which were found in the same stocks. So these pistols could be sold lucratively for "hard" foreign currency.

Illustration 15: *Czech proof mark "rampant lion"*

Walther P.38 and Mod. HP Austria Post-1945

By the end of World War Two there were large amounts of weapons on Austrian soil, too. At first the Austrians did not have direct access to these weapons, however, as the Allied Powers kept them under lock and key. From 1945 up to 1952 Austria had no own armed forces, but only a "Gendarmerie", later called "Bundesgendarmerie" (roughly translated: constabulary or civil police). These were armed with former military weapons, e.g. with P.38 pistols.

In May 1955 the Austrian "Bundesheer" (federal army) was founded, which included the air force ("Luftstreitkräfte"). The often quoted Austrian navy ceased to exist in 1918, when Austria was cut off from the Adriatic Sea through the loss of territory and thus of the harbours Triest (now Italy) and Pola (now Croatia).

The "Bundesheer" was first armed with P.38 and Colt M1911A1 pistols.

Early in 1956 all available P.38s from the army, police, customs and other authorities were collected and submitted to the "Bundesheer" for a general inspection.

As not enough pistols suitable for service were then available, the Austrian government appealed to the Allied Powers to hand them out all the captured weapons – including P.38 pistols. This request was more or less agreed to by the Allied Powers. So in 1957/58 Austria was in the possession of about 8000 P.38 pistols. Those which were unsuitable for use were cannibalized. Suitable ones were reconditioned extensively. Hereby old parts, but also new parts, made by Walther/Ulm, were used in this refurbishing program.

Illustration 16: *Austrian markings on P.38 pistols [left: Federal Austrian Eagle; right: "Bundesheer"]*

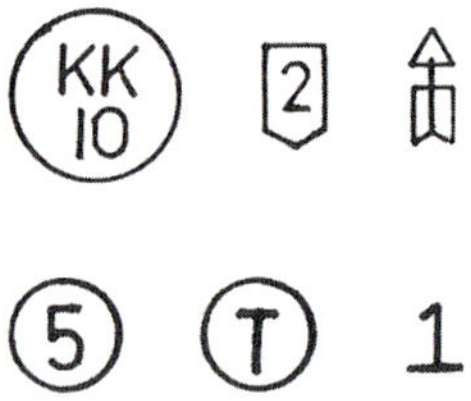

Illustration 17: *Possible markings on replacement barrels for Austrian P.38 pistols*

Illustration 18: *Commercial proof mark of Vienna proof house*

Refurbished P.38 and Mod. HP pistols

Remarks:

The Austrian "Bundesheer" obviously procured, refurbished and then issued or put in storage only P.38 and Mod. HP pistols – no Mod. P.38 pistols.

Characteristics:

- Slides, barrels and most of the frames from pre-1945 production
- Minor parts (firing pins, safeties, and some grip plates) from Walther/Ulm or from cannibalized P.38s
- Finish re-blued or touched up – mostly including the locking blocks and the sears
- Original markings were not grinded out (except most slide legends on the left hand side of Mod. HP slides)
- Serial numbers, which did not match were overstamped with the one taken from the frame. Original serial numbers on the slide were either milled or grinded off or otherwise defaced. The suffix resp. prefix letters were not always transferred to the new serial numbers.
- Subsequent stamping with "BH" (for "Bundesheer") and the Austrian Federal Eagle – see illustration 16

Some of the refurbished P.38 pistols have barrels, which came from a so far unknown source. These barrels show stamps, which are not decoded yet: "KK 10" (in a circle) and other signs – see illustration 17. It is very likely that these barrels were manufactured in "Plant 10", a military arsenal and factory in Kasanlik/Bulgaria.

When refurbishing pre-1945 P.38 pistols for the "Bundesheer" new parts from Walther/Ulm manufacture were increasingly used. Later the Austrian procurement office even purchased complete Walther/Ulm P38/P1 pistols with lightweight frames in order to complete the existing depots of P.38s for the "Bundesheer".

To distinguish the newly purchased P.38s from the refurbished ones the first ones were designated "P.38 alter Art" or "P.38 a.A." (P.38 old type).

There were about 300 to 500 Mod. HP pistols in post-1945 Austrian service. During the reconditioning process the Mod. HP slide legends on both sides, the serial number on the frame and the model designation were mostly removed. Afterwards the serial number was transferred to the left hand side of the slide and frame to make it visually the same as on P.38 pistols. If the serial numbers of frame and slide did not match, the number of the frame was retained – with exceptions though as usual.

Subsequently these pistols were refinished with a blue-black finish – including sear and locking block. There were also exceptions.

By the end of 1960 these refurbished P.38 and Mod. HP pistols were either issued to the "Bundesheer" or put in storage.

From 1986: The P.38, Mod. HP resp. P38/P1 pistols were gradually taken out of service and sold on the commercial market. So these pistols needed commercial proof marks. The proofing was done by the Vienna proof house, which stamped these pistols with "NPv" (for nitro proof Vienna) – see illustration 18. Next to the proof mark there is the year code (e.g. GTM" for July 1985) – see appendix. These commercial proof marks were applied on the right hand side of the frame and slide.

Walther P.38 Israel Post-1945

Israel was one of the "third countries", which procured P.38 pistols from the Czech Republic (see above). Already in 1948 the Israeli liberation movement "Haganah" purchased P.38 pistols from the French Mauser or Czech production.

In 1949/50 after the new state Israel and the "Israel Defense Forces" (IDF) had been founded, most of these pistols were transferred to the arsenals of the IDF. But there were only a few P.38 pistols available. So these were issued to non-commissioned officers and officers only. Before these pistols were taken over by the IDF, they were inspected and if necessary refurbished or repaired. No newly manufactured parts were used for this repair work, however, but parts from cannibalized pistols.

Illustration 19: *Israeli proof and property marks [left hand side: stamp "nun" (Israel proof mark); middle: stamp "tsadi" (property mark of the IDF); right hand side: stamp "mem" (property mark of the Israeli police)]*

P.38 Pistols for the Israel Defense Forces (IDF)

Characteristics: as P.38 pistols from World War Two production, but

- Marking with the Hebrew letter "tsadi" (first letter of the word "tsahal" = Israel Defense Forces) as a property mark – see illustration 19
- Israeli proof mark "nun" – see illustration 19
- These markings are mostly positioned inside a circle or a Star of David

P.38 Pistols for the Israeli Police

Characteristics: as before, but

- Marking with the Hebrew letter "mem" (first letter of the word "mishtara" = Israeli police) – see illustration 19

Before they were used as service pistols the P.38 pistols for Israel underwent a refurbishing process, although most of them had already been reworked before by their previous user. Especially some of the police service pistols went through several of these processes. So mostly the condition of these pistols is accordingly: heavily used, no more matching numbers, rounded edges by several polishing processes etc.

There were no new Walther/Ulm P38/P1 procured.

In the 1950s Israel sold many of these P.38 pistols either to Third World countries or on the commercial market. Most of these pistols were then refurbished again by semi-state owned companies like "Codil Ltd. Israel" (The latter stamped the pistols before sales with the letter "C" or "Codil Ltd. Israel").

Walther P.38 Turkey Post-1945

In Turkey there were never any Walther P.38 pistols produced under license or as unlicensed copies – in contrast to the PP copies made in Kirikkale (see there).

From 1950 to 1955 the Turkish procured some P.38s to be tested in troop trials and later by the police as well. These trials were controlled by the semi-state owned institution "Makina ve Kimya Endüstrisi" ("MKE"), which had its headquarters in Ankara and by the government arsenal Kirikkale.The latter is the place where the Walther PP copies were produced from 1947 to 1948.

Later a very small series (only five specimens?) of P.38 type pistols was manufactured – again unlicensed. It is not known whether used or newly made parts were used in this production.

P.38 Turkish Production

Characteristics: as P.38 pistols from World War Two production, but

- Slide legend on left hand side: "Kirikkale Tüfek Fb. Cap 9 mm" (for: Kirikkale weapons factory cal. 9 mm)
- Slide legend on right hand side: "MKE" over the year of production
- Some pistols with an additional legend "T.C.M." (for: "Turkiye Cumhurieyeti Mali", property of the Turkish Republic)
- Smooth wooden grip plates
- Serial number range (according to Warren Buxton): 5001 – 5005

Walther P38 and P1 Walther/Ulm Post-1945

Preliminary remarks:

The new model designation was: "P38" (without a dot) instead of "P.38" (with a dot).

The post-World War Two manufactured pistols are dealt with as late as here only because the ones described before were basically either refurbished pre-1945 P.38 pistols or post-war copies. The actual further development of this model was done by Walther/Ulm.

After World War Two Germany was subdivided into four different zones by the Allied Powers. Within these four zones there were strict limitations as to the manufacture and use of weapons as well as to the formation of armed and centralized police forces. The Allied Powers agreement was differently handled and observed by the four allies though – cf. the production of Mauser P.38 pistols under French control or the "Pistole 1001" and P.38 pistol production under Soviet control (see there).

As described in the chapter "P.38 GDR" the first law enforcement forces in the four different zones were more or less decentralized police forces. They were armed with a mishmash of service weapons – some of them surplus or captured weapons. Here the carbine 98k or the Walther pistols PP, PPK and P.38 played an important role, as most of the new law enforcement officers were familiar with these weapons due to their former service in the "Wehrmacht".

The American zone was an exception: Here all seized weapons of German origin were destroyed soon after 1945. Thus American weapons were issued almost exclusively to the new police there: M1 carbines, Colt Mod. 1911 and 1911-A1 as well as Smith & Wesson revolvers.

After the three zones of the Western allies were merged to become the new Federal German Republic in 1949, the new West German "Bundeswehr" (federal military force) was founded in May 1955. This new army was first armed with a hodgepodge of weapons, too – e.g. the British Enfield rifle No. 4 Mk1. The military first used some pre-1945 made P.38 pistols, but mostly foreign pistol models as service pistols, whereas the police forces preferred the Walther PP and PPK pistols. Some P.38s were used by the police as well, however.

The police P.38s were reworked – often making one out of two pistols. Swastikas were defaced, the finish was reblued and the serial numbers of the major parts were matched with the ones on the frames. Old, original serial numbers were grinded out or overstamped.

The first police P.38s were marked with a stylized police star (similar to the one used in East Germany) – often with the heraldic animal or the coat-of-arms of the city or district in the center.

From 1957/58 the police forces were increasingly equipped with the new Walther/Ulm P38 as a service pistol. An exception here was the West Berlin police. Due to the Four Powers Agreement they were allowed to purchase only weapons of non-German origin. So the West Berlin police adopted the "French" Manurhin made "Pistolet P1" (see there), which actually was produced by Walther/Ulm.

Up to 1986 all Walther PPs and PPKs were made by Manurhin – no matter what the slide legend says – whereas Manurhin never produced any P38/P1 pistols, even if the slide legend carries their company address.

From 1955 the standard "Bundeswehr" ("BW") sidearms were first the Colt pistols

M1911 (BW designation "P51) and the M1911A1 (BW designation "P52"). When parts of the German "Bundesgrenzschutz" (border police) were integrated into the "Bundeswehr" some Swiss pistols SIG 210-1 (BW designation "P2") and some Spanish Astras (BW designation "P3") were transferred to the BW arsenals.

From 1957 onwards the "Bundeswehr" received new Walther/Ulm P38 pistols. These were marked "P38" on the left hand side of the slide. The right hand side of the slide carried the new NATO supply number: 1005-12-120-6168.

From 1963 the BW was delivered the Walther/Ulm P38 pistols with the new model designation "P1" on the left hand side of the slide, as "P1" was the military designation of this Walther pistol. The first shipment came with steel frames. The standard P1 pistols have lightweight alloy frames though. So the "Bundeswehr" later mostly replaced these steel frames by alloy frames.

The Walther PPK was adopted by the BW as pistol "P21" – see there.

The supply contract for the "Bundeswehr" P38/P1 pistols was placed without a public invitation to tender. To fulfil this contract Walther/Ulm even had to ask for a credit, which was covered by bank guarantees of the federal state Baden-Württemberg and by the federal government.

Compiling a chronology of the P38/P1 production is almost as difficult as compiling the one of the pre-1945 production. Here, too, there are hardly any reliable and congruent production data as to the technical changes in production.

The files of Walther/Ulm deal with the sales rather than the production data. Those differ from the data given in the "Bundeswehr" service regulations and manuals. Some of these again differ from the data given in the Walther literature.

Manurhin is obviously very reluctant to release any information. In 1984 Manurhin was taken over by the French Matra group. Manurhin today produces mainly machinery for the production of ammunition. So the pistol production is more or less history for them.

According to French laws the documented production data must be stored by the district government ("préfecture"). But there they are under lock and key.

If the comparison of data cannot lead to reliable results, the relevant data are given with the remark "about" or "approximately" in the following text.

The P38 and the P1 pistols are discussed together here, as they were produced within the same time frame, in the same serial number ranges and with identical mechanical or production changes.

From 1963 the pistol "P1" was officially the military or police variation of the commercial P38 pistol, i.e. it was available to "Bundeswehr", municipal and state police forces, border police, customs and other authorities. This differentiation was not observed consistently though: If no P1 pistols were available for a shipment to any authority or government agency, P38 pistols were delivered.

The P1 pistols for the "Bundeswehr" and the "Bundesministerium des Innern" (Ministry of the Interior – in charge of e.g. border police and state police) were checked and proofed by the German "Güteprüfdienst des Bundesverteidigungsministeriums" (quality inspectorate of the Ministry of Defense). They were marked with their acceptance stamp (often referred to as "Eagle with Drooping Wings"), which included the identification number of the inspector. This stamp is both acceptance stamp and proof mark – see appendix.

In cooperative assistance the German "Güteprüfdienst des Bundesverteidigungsministeriums" (quality inspectorate of the Ministry of Defense) checked and proofed weapons from other contracts as well, e.g. the "Personal Protection Weapons" for the British army and police.

Walther P38/P1 Walther/Ulm

SN range: **001 001 - 473 201**

Production: 5/1957 - 1997
Calibers: 9 mm Luger (9 x 19 mm) - 8 rounds
7.65 mm Parabellum = .30 Luger - 8 rounds

(Pistols in this caliber were mostly delivered to countries, where the possession of pistols in the "military" caliber 9 mm Luger was forbidden - e.g. Italy. Conversion kits as well as complete pistols were produced in this caliber, the latter from 1968 from SN 313 000. This production ceased on 2nd January 1991 though)

. 22 l.r. (complete pistols or conversion kits)

Total Length: 8.5 in.
Barrel Length: 4.9 in.
Weight: 33.4 oz. (steel frame) or 27.5 lightweight alloy frame
Rifling: 6 grooves, right-twisted

Operation: locked blowback breech with firing pin and exposed hammer; pivoting locking block (below the barrel), double action/ single action trigger mechanism

Characteristics: as pre-1945 made P.38 pistols, but

- Slide legend on left hand side: "Carl Walther Waffenfabrik Ulm/Do. P38 / Cal. 9 mm" resp. " ... P1" or "P38 cal. 7,65 mm Para" with Walther banner – mostly in two lines, rarely in one line (examples for one line slide legend: SN 019 053 – 020 756, SN 280 361 – 302 468, SN 308 442 – 309 961 – in the early 1960s); from 1975: commercial P38s without caliber designations
- Serial numbers: With early P38 pistols the serial number was applied to the left hand side of the frame. If these pistols were used as military or police service pistols, the last three digits were often subsequently stamped onto the slide, the barrel and the locking block. With later P38/P1 pistols the full serial number was applied to the frame and the slide, whereas the last three digits were stamped onto the slide, the barrel and the locking block.
- Frames: very first frames made of steel; standard frames made of blue-black anodized lightweight alloy; steel frames available as an optional from the end of 1986 again
- First frames not reinforced by a hexagonal cross-pin yet, but with a hump just below the trigger axle (cf. P.38 pistols from April 1943 – see there).
- Freestanding barrel held in a sliding barrel carrier; barrel (with its rifling) and barrel carrier are no longer a unit made of one single forged block of steel without a pressed-in liner (cf. pre-1945 made P.38 pistols), but now have a pressed-in liner.
- Three types of liners were used:

05/57 - 09/72:	liner is secured by a cross-pin below the chamber
10/72 - 10/74:	liner is secured by a narrow ring at the rear of the chamber
11/74 - 02/92:	as before, but with a wider ring
from 02/92:	liner omitted - i.e. carrier and barrel are a unit made of one single forged block of steel without a pressed-in liner again (cf. pre-1945 made P.38 pistols); only very few pistols with this feature were acquired by "Bundeswehr", however, as from 1992 the P1 was gradually taken out of service to be replaced by the Heckler & Koch pistol "P8".

- Metal finish: as a rule matte-black phosphated finish, but for commercial sales a high-gloss polished finish was available

Picture 59: *Walther P38, commercial variation, SN 473 160, caliber 9 mm Luger, produced by Walther/Ulm in 1997, one of the last ones produced in Ulm*

as an option – see below

- Slide bridge: slightly more angularly shaped as on pre-1945 made P.38s
- Very early P38 pistols with a milled rail on either side of the slide top cover to protect this cover
- Firing mechanism modified: firing pin no longer flat, but round; first unpolished, later polished
- Slide: a little cutout in the front of the inside to make disassembly easier; i.e. slide has to be retracted only about 10 mm to be able to rotate the disassembly lever down
- Locking block: 5/1957 – 2/1972: locking block made using precision casting with a cast-in hardening mark (combination of a letter and/or a number in a square); from 2/1972: locking blocks are forged, hardening mark omitted
- Slide with a short, half round extractor relief slot similar to that used on P.38 pistols from April 1943 – see there
- Rear of slide: not two separately milled out areas for the signal pin and the hammer (cf. early P.38 pistols), but the two millings are merged (cf. P.38 pistols from April 1940)
- Slide serrations on both sides of slide: first 16, later 27 serrations
- Grip plates: two-piece made of black, checkered plastic ("Novodur" – a thermoformed ABS plastic made by Bayer) with a little thumb rest, which is similar to the one found with Mod. HP and early Zero Series P.38 pistols; wooden grip plates were available, too
- Magazines for commercial P38 pistols: on left hand side of body stamped with "P38 9 mm" and Walther banner; magazines for P.38/P1 service pistols: on left hand side of body stamped with "P38 9 mm" resp. "P1 9 mm" and Walther banner, on spine stamped with the date (e.g. "9/80") and the acceptance stamp of the "Güteprüfdienst" and the NATO supply number "1005-12-120-0892" (the latter omitted from 3/1976)
- Proof marks on commercial P38 pistols: Ulm proof house stamp ("Federal Eagle

Table: Serial Numbers Walther P38/P1 Ulm

Model	Variation and Serial Number Range	Production Period
P38/P1, caliber 9 mm Luger	**lightweight alloy frame**	
P38	001 001 E – 024 021 E	05/57 – 07/68 (some of them with polished finish)
P1 or P38	001 001 – 299 707	05/57 – 07/66
P38 (no P1s)	299 708 – 370 000	03/67 – 10/75
P1 or P38	370 001 – 390 000	02/76 – 12/77
P38 (no P1s)	390 001 – 399 999	09/76 – 08/78
P1 (no P38s)	400 000 – 410 000	06/77 – 09/78
P1 and P38 (serial number range assigned in duplicate! – for P38 as well as for P1 pistols)	410 001 – 410 800	09/78 – 05/79
P1 or P38	410 801 – 473 201	05/79 - 1992 (some of them with steel frames)
P38, caliber 9 mm Luger	**steel frame (a total of about 1,540 P38s were made with steel frames)**	
P38 (no P1)	008 250 – 008 448	06/57 – 07/68
P38 (jubilee edition "100 Jahre Walther")	505 001 – 505 713	12/86 – 04/89 (7123 pistols, about 60 of these with a light engraving; SN 505 713 in caliber 7.65 mm Para)
P38 (jubilee edition "50 Jahre P38")	001 – 500	1988 (only 289 pistols made! – 257 pistols of these in 9 mm Luger, 32 pistols in 7,65 mm Para)
P38 (with longer barrel)	001 – 073 with lightweight or steel frames – some pistols with polished finish	1978 - 1989 (73 pistols, 41 of these with 6" barrel, 32 with 7" barrel)
P38, caliber 9 mm Luger	**polished finish with lightweight or steel frames**	
P38	no separate SN-range	1957 – 12/72
P38, caliber 7,65 Para	**complete pistols with lightweight frames or conversion kits**	
P38	from about SN 313 000 (but no separate SN-range)	2/1968 – 1988 about 5000 pistols
P38 conversion kits caliber .22 l.r.	**with insert barrels**	
P38 or P1	0001 – 8313	1963 – ca. 1970
P38 conversion kits caliber .22 l.r.	**with separate barrels**	
P38	0001 – 8313	1970 - 1974
P38, caliber 22 l.r.	**as complete pistols**	
P38	0001 – 8313 and from 300 000	1967 – 1987 about 5000 pistols
P38 SD	**with silencer**	
P38	no separate SN range	?

over N" with "antler", from 1961 with the end digits of the year, from 1977 with the year code – see appendix); proof marks on P38/P1 service pistols: acceptance stamps of the "Güteprüfdienst des Bundesverteidigungsministeriums" (quality inspectorate of the Ministry of Defense) both on the frame and the slide ("Eagle with Drooping Wings" with identification number of inspector) – see appendix

- Slide legend usually in two lines, rarely in one line – e.g. in the early 1960s
- Variation for exports to "problem countries" (e.g. South Africa): one line slide legend, no proof or acceptance marks, no date stamp

Walther P38 and P1 Walther/ Ulm for German "Bundeswehr"

The very first shipment of 59 800 Walther service pistols (still with "P38" model designation – see above) to the "Bundeswehr" was made on 16th August 1956 – beginning with serial number 001 001.

From June 1957 up to September 1963 the "Bundeswehr" received the major shipments of Walther P38 service pistols.

Surprisingly the approval requirement for the first P38 service pistols was to survive a total of 500 rounds only. This limit was later raised to 3000 rounds and then 10 000 rounds.

The military model designation "P1" was used in the slide legend from October/ November 1963 onwards only.

Before 1945 the Mod. HP and the Mod. P.38 pistols were the commercial variations of the Walther P.38 service pistol. In post-1945 times the P38 was the commercial variation, whereas the P1 was the service pistol – with exceptions (see above).

The full, official military designation of the "Bundeswehr" P1 was "Pistole, 9 mm x 19, P1". The manufacture, the proof, the acceptance, the marking and even the packing was performed according to the strict regulation "Technische Lieferbedingungen" ("TL" – technical delivery conditions) of the "Bundesamt für Wehrtechnik und Beschaffung" (BWB – federal authority for defense technology and procurement). This was supervised by the inspectors of the "Güteprüfdienst des Bundesverteidigungsministeriums" (quality inspectorate of the Ministry of Defense). They checked the materials and the interchangeability of the parts. This was followed by a proof firing using a high pressure cartridge and a test firing for accuracy.

The maintenance of the P38/ P1 service pistols by the military units (company, resp. battalion) and the two "Systeminstandsetzungszentren" (maintenance centers) in St. Wendel resp. in Doberlug-Kirchhain was strictly regulated by the "Leistungsbeschreibung Instandsetzung LB INST für Pistole P1" (service regulations for maintenance …).

In 2000, i.e. before the official decommissioning of the P38/P1 as a service pistol, the "Bundeswehr" still possessed a total of 120 000 pistols – although from 1992 some pistols had already been taken out of service.

Characteristics: as commercial P38s, but

- Production period: 10/1963 - 1992
- Slide legend from October 1963 on left hand side of slide changed to "Carl Walther Waffenfabrik Ulm/Do. / P1 Cal. 9 mm" and from 07/1968: "P1 Kal. 9 mm", with date of manufacture (e.g. "4/80") and

Picture 60: *Walther P38, "Bundeswehr" variation, SN 022 337, caliber 9 mm Luger, produced by Walther/Ulm in 11/1957, NATO supply number on right hand side of slide, one of the first P38 pistols delivered to the "Bundeswehr", frame and slide still unreinforced*

last three digits of serial number – see illustration 20

- Acceptance/proof marks: acceptance stamps of the "Güteprüfdienst des Bundesverteidigungsministeriums" (quality inspectorate of the Ministry of Defense) ("Eagle with Drooping Wings" with identification number of inspector) on left hand side of frame, right hand side of slide, front of barrel carrier and on locking block
- Serial number application: full SN on left hand side of frame, last three digits on front of barrel carrier and on locking block; from 11/1961 last three digits on left hand side of slide, too
- NATO supply number: up to 11/1963 on right hand side of slide (1005-12-120-6168) and on spine of magazine (1005-12-120-0892), (the latter omitted from 3/1976)
- Property marks of the "Bundeswehr":
- From 1963: "BW";
- From 2/1977: "Bw" (with a small "w" to prevent confusions with the abbreviation for the federal state of Baden-Württemberg);
- From 1981: 1956 -1963 made P38 pistols were often subsequently stamped with "Bw" during maintenance
- Stamp "dot within a circle" (see appendix) on left hand side of slide and frame: limiting measurements and fittings comply with the specifications of "Bundeswehr" and other authorities.
- Markings for maintenance by "Systeminstandsetzungszentren" (maintenance centers) in St. Wendel resp. in Doberlug-Kirchhain – e.g. on left hand side of frame: "162750W6.92" for order number, sign for St. Wendel, date of maintenance
- SN range: 230 000 – 467 070 (about 200 000 pistols)

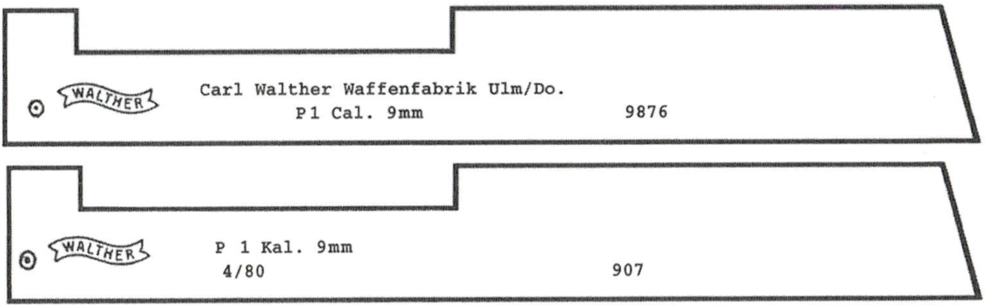

Illustration 20: *slide legends on Walther P1 pistols [inserts from top to bottom: example of a legend 10/1963 – 06/1968; example of a typical legend from 07/1968 onwards]*

Walther P38 and P1 Walther/Ulm for German Authorities ("BMI")

The Walther P1 pistol was introduced by the "Bundesministerium des Innern" ("BMI" – Ministry of the Interior), too. The BMI procured weapons for the "Bundesgrenzschutz" ("BGS" – border police) and the riot police forces ("Bereitschaftspolizeien") of the federal states.

In May 1962 these authorities issued common service regulations for the Walther P38/P1 pistols ("PDV Nr. 903").

In January 1962 the BGS introduced the Walther P38/P1 officially as a service pistol and took this pistol model out of service again latest in 1982.

Characteristics: as the "Bundeswehr" P38/P1 pistols, but

- Marking "BMI" on right hand side of frame – at least on most pistols
- Most pistols without the NATO supply number (NSN)
- Acceptance stamps of the "Güteprüfdienst des Bundesverteidigungsministeriums" (quality inspectorate of the Ministry of Defense) ("Eagle with Drooping Wings" with identification number of inspector) on left hand side of frame
- For shooting practice the BGS and the riot police forces acquired conversion kits in caliber .22 l.r. These were issued in wooden boxes, zeroed-in for a certain service pistol, and serial numbered accordingly. These kits were rarely used, however – see below, too.

Walther P38 and P1 Walther/ Ulm Development and Technical Changes

Picture 61: *Walther P1, "Bundeswehr" variation, SN 385 378, caliber 9 mm Luger, produced by Walther/Ulm in 2/1977, frame and slide still reinforced, phosphated finish*

Most of the technical changes described below were adopted more or less simultaneously for both P38 and P1 pistols:

- 10/1963: slide legend on left hand side changed from "… P38 Cal. 9 mm" to "… P1 Cal. 9 mm"
- 02/1968 (from about SN 310 000): introduction of reinforced slide, i.e.

a) wall thickness increased from 2.5 mm to 3.2 mm (cf. "Technische Anweisung für Material- und Formänderung (TAM-F)" - from 15th February 1968), hence often named "fat" slide in US literature

b) number of slide serrations at the rear of the slide: increased from 16 to 27, serrated area now extends to the end of the safety lever when in position "fire"

c) slide stamped stamped with a "four pointed star" on left hand side below the caliber designation, which indicates "Form geändert", i.e. design modified - see appendix

d) contour of the ejection opening in the slide is cut less far down, i.e. providing a straight line with the upper edge of the slide, which simplifies manufacture

e) take down lever reinforced, stamped with a "four pointed star", which indicates "Form geändert", i.e. design modified

f) slide legend changed again from "P1 Cal. 9 mm" to " ... Kal. 9 mm" - cf. "Bundeswehr" P21 pistols - see there.

g) sights modified: front sight wider and with a contrasting, white dot; rear sight with a 4 mm cutout and contrasting white paint (only a 2.3 mm cutout without contrasting paint before)

h) Some pistols with reinforced slides are marked "P38/II".

- 04/1971: introduction of spring-loaded ejector
- 02/1972: introduction of forged locking block (instead of the ones made using precision casting), hence the cast-in hardening mark was omitted
- 09/1972: barrel liner with a narrow ring at the rear end
- 10/1974: barrel liner with a wider ring at the rear end
- 02/1975: sights modified again – i.e. front sight wider and marked with a number (2, 3 or 4) for the height of the front sight; contrasting, white dot (standard from now on – for Walther P4 pistols, too);
- Reinforced rear sight; marked on the front with letter "L" (for left) or "R" (for right), letters indicate the lateral offset of the cutout (0.3 mm to either side); contrasting white bar;
- 06/1975: introduction of reinforced frame, i.e. hexagonal, hardened cross-pin in frame (just in front of the trigger axle); pin serves as an impact surface for the locking block (standard from now on – for Walther P4 and P5 pistols, too)
- 03/1976: NATO supply number on spine of magazine omitted
- 02/1977: property mark changed from "BW" to "Bw"
- 11/1977: take down lever modified again, i.e. lever thumb spur angled about 15° to provide a better grip for the thumb;
- Introduction of polished firing pin
- 07/1983: introduction of new firing pin, reinforced by a bigger bulge at the rear end
- 02/1992: introduction of new barrel/barrel carrier unit without a pressed-in liner; this modification was still accepted by the "Bundeswehr", but hardly any P1 pistols were procured with this design feature any more, as the P1 was gradually taken out of service
- 07/1992: contour of magazine lips and hardening altered

Remarks:

The technical changes mentioned above were not always adopted with the dates given here. Mostly existing parts from stocks were used up in production first.

Maintenance staff and "Bundeswehr" inspectors often refitted older P38s/P1s with newly made parts. So hardly any P.38/P1 pistol is still in the original condition as it was once issued.

Walther P38 and P1 Walther/Ulm Special Variations

a) Walther P.38 export models (5/1957 – 7/1968)

Characteristics: as standard P38 pistols, but

- Serial number suffixed by letter "E"
- P38 for export to the USA often with the additional legend "P.38/II"
- Some pistols with a high-gloss polished finish
- Not all of the P.38 pistols with the suffix "E" went into export though. Some pistols were used to fulfil other contracts, e.g. for the "Bundesgrenzschutz" (BGS)
- SN range: 001 001E – 024 021E

b) Walther P.38 Pistols with polished finish (5/1957 – 12/1972)

Characteristics: as standard P38 pistols, but

- Slide and barrel group: high-gloss polished finish, instead of matte phosphated finish
- No specific SN range

c) Walther P38 Pistols with steel frames (5/1975 – 4/1989)

- See jubilee editions below, too
- SN range in 9 mm Luger:
- 008 250 (6/57) – 008 448 (7/68) and
- 505 001 (12/86) – 505 713 (4/89)

Characteristics: as standard P38 pistols, but

- Steel frame (similar to pre-1945 Walther P.38s)
- Some pistols with engravings or high-gloss polished finish
- SN 505 001 to 505 713: with additional slide legend "100 Jahre Walther"
- Wooden grip plates optional
- Only one pistol known in caliber 7.65 mm Para (SN 505 713)

d) Walther P38 "L" – with long barrel (1978 – 1989)

Characteristics: as standard P38 pistols, but

- Most pistols with steel frames
- Barrel length: 6" (= 152 mm, 41 pistols made) or 7" (= 178 mm, 32 pistols made)
- SN range: 001 – 073

e) Walther P38 SD (with silencer)

Remarks:

Ten to twelve P38 pistols are known, which were equipped with a fixed silencer with matching numbers. These pistols were made for a customer in Saudi Arabia, but never issued. In the 1970s they were auctioned in Great Britain.

These P38 SD pistols are not identical with the pre-1945 ones, which were rebuilt in London in the 1950s by the renowned gunsmith Gogswell & Harrison. The British SAS (Special Air Service) is said to have had some pre-1945 P.38 pistols reworked to P38 SD pistols in the 1990s, too.

One fact is for sure: Seven P38 SD pistols (SN range around 307 260) were produced by Walther/Ulm. In July 1969 six of these were delivered to Denmark.

Characteristics: as standard P38 pistols, but

- Slide hold-open lever is replaced by a slide locking lever on left hand side of frame to make sure that the slide does not make any unnecessary noise when repeating
- Early variation: silencer is screwed onto the barrel;

Picture 62: *Walther P38 SD, SN 307 267, caliber 9 mm Luger, produced by Walther/Ulm in 1968*

- Later variation: silencer is equipped with a mounting clamp to be slid over the barrel and to be locked in place with a special tool
- Front sight cannot be used, so the silencer has its own set of sights
- Delivery in a special long box (which was used as well for P38s with long barrels) labelled with "Pistole P38 SD"
- P38 SD pistols were issued amongst others to the "Fernspäher" (special forces) of the German "Bundeswehr".
- No separate SN range

f) Walther P38 in caliber 7.65 mm Parabellum (2/1968 – 1988)

Characteristics: as standard P38 pistols, but

- Available as complete pistols or conversion kits
- Lightweight frames – cf. jubilee P38 pistols "100 Jahre Walther"
- About 5000 pistols made
- SN from about 313 000, but no separate SN range

g) Walther P38 "N"

Characteristics: as standard P38 pistols before October 1974, but

- Slide legend on left hand side: "Carl Walther Waffenfabrik Ulm/Do. P38 / Cal. 9 mm"
- Unreinforced frame
- Reinforced slide
- Locking block forged
- Barrel liner with narrow ring
- The following characteristics differ from the standard P38 of that period:
- Pivoting lever on left hand side of slide without safety function, but with de-cocking function only – cf. Walther P38k or P4 pistols
- Metal finish: neither blued nor anodized, but coated with a black lacquer
- Additional letter "N" hand-stamped behind serial number

- Full serial number on right hand side (!) of slide
- Three final digits of SN hand-stamped on front of barrel carrier and on frame
- "WSD" on right hand side of slide above the serial number
- SN: 358 677 (only one specimen known?)

It is known, which reseller this P38 pistols was delivered to, but it is unknown, however, whether this P38 pistol is part of a small series.

h) Walther P38 caliber .22 l.r. (1963 – 1987)

SN range: 001 - 8313 and again **from 300 000**

Remarks:

The .22 l.r. P38 pistols were produced as complete pistols as well as conversion kits in the same SN range (SN 0001 – 8313). Very early conversion kits carried no serial numbers yet.

All in all there were about 5300 complete pistols or conversion kits produced in this caliber. Most of them were used for shooting practice by the BGS and the riot police forces of the federal states. Hence they were marked "BMI" – see appendix. The German "Bundeswehr" purchased only very limited amounts for test purposes.

1) Conversion Kits with insert barrel (1963 – 1970)

Characteristics: as standard P38 pistols, but

- Kit consists of slide (early, unreinforced variation), insert barrel and .22 l.r. magazine
- Inside of slide and slide bridge: some parts milled away to reduce the weight from 9.5 oz. to 7.7 oz. (1959 prototype still without the milling at the slide bridge)
- Some of the slides polished in the slide legend area
- Some of the slides without a slide legend, only the box is labelled – e.g. with "PD Berlin" (for "Polizeidirektion Berlin")
- Firing pin modified to fire the rimfire cartridge
- Signal pin (loaded chamber indicator) omitted because of the rimfire cartridge
- Only one recoil spring due to the weaker recoil, only one lateral cam inside the slide for the recoil spring

2) Conversion Kits with separate barrel (1970 – 1974)

Characteristics: as before, but

- Kit consists of slide (early, unreinforced variation with millings), barrel carrier with pressed-in liner in caliber .22 l.r. and .22 l.r. magazine
- The insert barrel was abolished, as this technical solution brought only problems, but no cost advantage with it.

3) Complete .22 l.r. P38 Pistols (1967 – 1987)

Characteristics: as before, but

- Slide (early, unreinforced variation with millings) and reinforced or unreinforced frame, barrel carrier with pressed-in liner in caliber .22 l.r. and .22 l.r. magazine
- No locking block, as one of the two lateral recoil springs takes over this "locking" function
- Inside of slide and slide bridge: some parts milled away to reduce the weight from 9.5 oz. to 7.7 oz. (1959 prototype still without the milling at the slide bridge)
- Firing pin modified to fire the rimfire cartridge
- No signal pin (loaded chamber indicator)
- Some P.38 pistols in caliber .22 l.r. for export with four digit serial numbers only

Picture 63: *Walther P38, SN 5369, caliber .22 l.r., produced by Walther/Ulm in 1987, frame reinforced, slide still unreinforced*

i) Walther P38 cutaway pistols

These pistols were mostly not made in the Walther factory, but by police, army or BGS armorers.

j) Walther P38 "S"

These pistols were not made by the Walther factory either, but by gunsmiths named Georg von Stavenhagen (died in 1978), Jürgen Ruhmann (died in 2014) and K. Grubert – as the "tuned" version of the standard P.38 pistol. Walther only provided five pistols as samples. These gunsmiths used parts from cannibalized P.38 or P38 pistols or newly manufactured parts from Walther/Ulm.

Characteristics: most pistols with modifications

- Modified sights, hammer spur, barrel, magazine release and safety lever
- Tuned and reworked trigger
- Trigger stop
- Wooden grips
- Some pistols with differing model designations on slide: "P38 S" (without Walther address and banner)
- Some pistols with a massive backstrap (similar to the Walther P5)

k) Walther P38 "Jubilee Editions"

Walther/Ulm produced not only a jubilee edition of the Mod. HP (see there), but also of the P.38 pistol:

1) "100 Jahre Walther" (1986) US Edition

Characteristics: as standard P38 pistols, but

- Slide legend on right hand side of slide: "P38 / 100 Jahre Walther (Walther banner) 1886 – 1986"
- Limited edition: 900 pistols

2) "100 Jahre Walther" (12/1986 – 4/1989) German Edition

Characteristics: as standard P38 pistols, but

- Slide legend on right hand side of slide: "1886 – 1986 / 100 Jahre" (Walther banner)
- Limited edition: actually limited to 500 pistols, but 713 were built
- Slides with engraving (oak leaves)
- Steel frames
- Smooth wooden grips
- SN range in 9 mm Luger:
- 505 001 (12/86)) – 505 712 (4/89)
- SN range in 7.65 mm Parabellum: 505 713 (4/89) – only one pistol made!

3) "50 Jahre P38" (1988)

Characteristics: as standard P38 pistols, but

- Slide legend on right hand side of slide: "P38 Nr. ..." (Walther banner)
- Slide legend on left hand side of slide: "50 Jahre P38 1938 – 1988"
- Limited edition: limited to 500 pistols, but only 289 pistols were made (257 in caliber 9 mm Luger and 32 in caliber 7.65 mm Parabellum)
- Steel frames
- Carved, wooden grips
- SN range: 001 - 289

Walther P38 and P1 Walther/Ulm Use by Foreign Armies and Police Forces

Remark:

There is only one contract known, where the purchaser did not simply order standard P38/P1 pistols or the export model:

Walther P38 and P1 Norwegian Army and Police

After World War Two the Norwegian army and police possessed a large number of pre-1945 made P.38 pistols – e.g. as a war booty.

In November 1952 there were 3800 P.38 pistols counted in Norwegian arsenals. Most of these pistols were still in a useable condition. However there was a great need for new barrels, slides and minor parts.

In a "cannibalizing campaign" in January 1953 there were 2300 P.38s in good condition assembled from parts which were obtained from these 3800 "useable" P.38 pistols.

In the 1960s Norway procured a large number of used or as new P.38s from the former GDR and the Czech Republic – e.g. "CZ46" pistols (see there). These pistols were either directly adopted as service pistols or used to obtain spare parts.

From 1957 to 1968 the Norwegian army additionally purchased parts from Walther/Ulm to repair or complete the service pistols.

After extensive tests the Walther P38 was officially adopted as the Norwegian military service pistol. Norway then purchased complete P38 pistols as well.

From 1963 to 1975 Walther/Ulm delivered a total of 25 000 P38 pistols to Norway – most of them to the Norwegian air force.

The following shipments could be spotted according to the relevant files, which are incomplete, however. The serial number blocks are not "closed" blocks either:

200 pistols:	SN 107 857 - 197 974
100 pistols:	SN 197 773 - 198 972
53 pistols:	SN 197 975 - 198 900
400 pistols:	SN 300 541 - 303 811
500 pistols:	SN 200 234 - 301 654
400 pistols:	SN 299 353 - 303 886
250 pistols:	SN 23 530E - 23 882E
140 pistols:	SN 347 220 - 347 360
55 pistols:	SN 352 826 - 352 881
1000 pistols:	SN 369 001 - 393 696

The Norwegian military did not purchase these pistols directly from Walther/Ulm, but from the Norwegian representative Egil With in Oslo. These pistols differed from the standard P38s as described below.

The Norwegian police had to be satisfied with P.38 or P38/P1 service pistols from old stocks. At least there are no purchases of new pistols known.

Characteristics: as standard P38 pistols, but

- Slide legend on left hand side: mostly with the additional Norwegian state emblem ("Crown over N" in a shield) positioned between the model and the caliber designation.

In October 1962 the Norwegian "Haerens Vapentekniske Korps" (roughly translated: weapon technological army corps) provided Walther/Ulm with a special technical drawing of this emblem.

This emblem was either roll-stamped between the model and the caliber designation or engraved on the right hand side of the caliber designation.

- A surprisingly large number of the P38 pistols shipped to Norway were taken from the "E" (for export) serial number range – see above.

Remarks:

A very detailed survey on the Norwegian P38 pistols can be found in: Per Mathisen, "Walthers P.38 Pistol in Norway", Mjoendalen, Norway 2005. The statements above are based on his research.

Walther P38 and P1 Portuguese Army and Police

After the Portuguese army had problems during their colonial wars with their Luger service pistol P08, they adopted the Walther P38/P1 as their new side arm for the army and the air force.

The shipments to Portugal were made from 1959 – some of them by Manurhin.

Already in about 1950 Portugal had procured a small batch of P.38 pistols – presumably from the Czech Republic. These pistols were made by Spreewerk/Grottau and were subsequently fitted with wooden grips. They were not used as normal service pistols though, but as ceremonial weapons for honor guards.

The Portuguese police acquired Walther P38/P1 service pistols, too. Amongst them about 1000 pistols for the "Guarda Nacional Republicana" (marked "GNR" on right hand side of slide).

But both the Portuguese military and the police P38/P1 pistols were standard pistols without a special slide legend.

Walther P38 and P1 for other foreign Army and Police Forces

Apart from Norway and Portugal other countries purchased standard Walther/Ulm P38/P1 pistols to be used as service pistols (in alphabetical order):

- Afghanistan: In 2006 the German Ministry of Defense delivered 10 000 P38 and P1 pistols for the military and the police.
- Argentina: a small amount for the army only

- Chad (West Africa): amount unknown
- Chile: army only
- Finland: for the Finnish UN troops (1965 – 1990)
- Lebanon: amount unknown
- Macedonia: amount unknown
- Mozambique: amount unknown
- Pakistan: amount unknown
- South Africa: police only (1976 – 1992)

Total production of Walther P38/P1 pistols: about 473,200 pistols

Walther P1 Manurhin ("Pistolet P1")

Picture 64: *"Pistolet P1" for the West Berlin police, SN 219 495, caliber 9 mm Luger, assembled by Manurhin in 1963, frame and slide unreinforced, "Berlin rosette" on left hand side of trigger guard*

SN range:	**from about 200 000 to about 450 000**
Production:	1963 - 1979
Calibers:	9 mm Luger (9 x 19 mm) - 8 rounds . 22 l.r. (conversion kits only)

Characteristics: as Walther/Ulm P38/P1 pistols, but

- Slide legend on left hand side: "MANUFACTURE DE MACHINES DU HAUT-RHIN / PISTOLET P1 / 9 mm x 19" (with Manurhin logo in the center and the addition "Made in France" or "France" only)
- French proof marks (mostly St. Etienne)

Illustration 21: *slide legend on Manurhin "Pistolet P1" pistols*

Remarks:

After Manurhin had manufactured PP and PPK pistols for Walther/Ulm from 1952, they started the assembly and sales of P1 pistols as "Pistolet P1" from 1963. This was done with parts which were provided by Walther/ Ulm in-the-white. These parts were stamped, blued, assembled and proofed by Manurhin. The serial number ranges are the ones used by Walther.

These P1 pistols with Manurhin slide legend were delivered to customers who were not allowed to purchase "German" weapons. This was e.g. true for the West Berlin police and customs authorities, which were forbidden by the Four Powers Agreement to buy "German" pistols. So they first acquired P.38 pistols made under French control by Mauser/ Oberndorf in 1945/46 – see there.

On 31st October 1963 the West Berlin police procured 14 630 "Pistolet P1". These were first issued to the "Bereitschaftspolizei" (riot police) and later to the regular police force, too.

These "Pistolet P1" were assembled from parts, which were actually intended for an export contract (presumably to Portugal). So some of the slides carry the marking "crossed cannons", which can be found on the berets of the Portuguese artillery, too.

The West Berlin P1 service pistols are marked with the "Berlin rosette" (a stylized police star) on the left hand side of the trigger guard.

Next to the P1 pistols the West Berlin police acquired Manurhin caliber .22 l.r. conversion kits for shooting practice (some of them with French proof marks).

In 1964 the West Berlin customs authorities procured 600 "Pistolet P1" and marked them with "BUND" on the right hand side of the slide.

If slides had to be replaced, this was done with later reinforced ones. These were then stamped with the West Berlin proof mark ("Berlin bear").

From 1982 resp. 1987 the West Berlin police and customs took these "Pistolet P1" out of service again.

Walther P38 Copies

Remarks:

Although the Walther P38/P1 pistols have been and still are popular pistols models, which were produced in large numbers, there are hardly any copies by other companies – in contrast to the Walther PP and PPK pistols (see there).

The British Webley & Scott pistol Mk1 and the Beretta 9/51 are said to have some technical features in common with the P38 model. It would, however, be a bit farfetched to classify these pistol models as P38 copies.

There are only two pistols, which can be regarded as P38 copies:

1) Erma EP 882 and EP 882S (Germany)

The company Erma was located in Erfurt/Thuringia from 1922 until 1945. There they developed, amongst other weapons, the submachine guns MP38 and MP40.

In 1951 Erma was re-established in Dachau/Bavaria and produced mainly alarm pistols.

They also produced a .22 l.r. copy of the P38 using die-cast zinc technology. This pistol shows the contours and many technical characteristics of the P38 pistol (e.g. freestanding barrel, blowback action with firing pin and exposed hammer, double action/single action trigger mechanism, take down lever, slide hold-open lever, pivoting safety lever in the slide, bottom magazine release in the heel of the grip). But there is no locking block and only one recoil spring (below the barrel). The barrel is fixed and only the rear part of the slide moves backwards when repeating. There is no real slide bridge either.

The Erma EP 882 comes with a short 5" barrel, whereas the EP 882S comes with 6" barrel.

The single-row magazine holds eight rounds in caliber .22 lr.

2) PHP MV (Croatia)

This pistol in caliber 9 mm Luger can be regarded as a Walther P38 copy in the wider sense, as it combines features of the Walther P38 and the Beretta 9/51. The "Prvi Hrvatski Pistolji" (first Croatian pistol) was manufactured in the 1990s in Ozalj/Croatia. It shares the following basic characteristics with the Walther P38: locked blowback breech with firing pin and exposed hammer; one massive, pivoting locking block; freestanding barrel held in a sliding barrel carrier; slide open on top, with a bridge at the front.

The PHP MV has only one recoil spring (below the barrel), an upper magazine release and the safety lever on the left hand side of the frame.

The double-row magazine holds 15 rounds in caliber 9 mm Luger.

For a short time the PHP MV was sold on the European and US commercial market – with little success, however.

Walther P38 K Walther/Ulm and Manurhin

Production:	2/1972 - 1981 (both Walther/Ulm and Manurhin)
Calibers:	9 mm Luger (9 x 19 mm) - 8 rounds 7.65 mm Parabellum = .30 Luger - 8 rounds (Pistols in this caliber were mostly delivered to countries, where the possession of pistols in the "military" caliber 9 mm Luger was forbidden - e.g. Italy)
Total Length:	6.3 in.
Barrel Length:	2.8 in.
Weight:	27.2 oz.
Rifling:	6 grooves, right-twisted
Operational:	locked blowback breech with firing pin and exposed hammer; pivoting locking block (below the barrel), double action/ single action trigger mechanism

Preliminary remarks:

It is doubtful whether there was a series production of the Walther P.38 "k" pistol before 1945. Presumably these were either single pieces or subsequently shortened standard P.38 pistols. So these P.38 "k" pistols mostly carry the front sight on the shortened barrel.

There were ten P.38s shortened in the workshops of the East German "Ministerium für Staatssicherheit" ("Ministry of State Security" of the former GDR, short: "Stasi") – see there.

Some P38 pistols were shortened by gunsmiths like G. von Stavenhagen (Germany) or John V. Martz (USA), but did not carry the designation P38 K. Martz even modified some of these pistols for the caliber .45 ACP.

Walther/Ulm seems to have had problems with the development of their P38 K pistols (here spelled with a capital "K"), as there were several experimental pistols and pre-series made before the actual series production could begin. Even within these pre-series there were technical changes and modifications.

In the serial number range 600 002 to 600 007 there were six P38 K pistols produced in the serial number range of the Walther P4, but with a "P38" slide legend.

Walther/Ulm P38 K 1st Pre-Series (1972)

SN range: 348 513 to 348 536 (24 pistols)

Characteristics: as Walther/Ulm P38/P1 pistols, but

- Slide legend on left hand side: "Carl Walther Waffenfabrik Ulm/Do. / P38" (with Walther banner, but without suffix "K")
- Proof marks: Ulm proof house ("Federal Eagle over N" with "antler" and year stamp "72")
- Short barrel
- Frame unreinforced made of lightweight alloy
- Front sight inserted in a dovetail on the slide bridge
- Slide reinforced, cover on top of slide – cf. P38 pistols
- Safety lever with safety and de-cocking function – cf. P.38 pistols
- These 23 shortened P38 pistols were issued to the "Polizeidirektionen" in Munich and Bremerhaven.

Picture 65: *Walther P38 K, SN 502 095, series production 2nd variation, caliber 9 mm Luger with 7.65 mm Parabellum conversion kit, produced by Walther/Ulm in 1978, frame and slide reinforced; fixed, milled front sight on slide bridge, adjustable rear sight*

Walther/Ulm P38 K 2nd Pre-Series (1973)

SN range: 351 001 to 351 025 (25 pistols)

Characteristics: as before, but

- Slide legend on left hand side: "Carl Walther Waffenfabrik Ulm/Do. / P38 K" (with Walther banner and with suffix "K")
- Proof marks: Ulm proof house ("Federal Eagle over N" with "antler" and year stamp "73")
- Safety lever with de-cocking function only – cf. later P4 pistols
- These 23 shortened P38 pistols were issued to the "Polizeidirektionen" in Munich and Bremerhaven.

Walther/Ulm P38 K 3rd Pre-Series (1973)

SN range: 355 031 to 355 044 (9 pistols)

Characteristics: as before, but

- No "closed" serial number block

Walther/Ulm P38 K 4th Pre-Series (1973)

SN range: 500 001 to about 500 100 (?)

Characteristics: as before, but

- Slide round on top, without a cover – cf. later P4 and P5 pistols
- Barrel carrier guided in slide with one lateral rail on either side (cf. later P5) and guided in frame with one lateral rail on either side – cf. P38 and P4 pistols
- Serial numbers known so far: 500 040 and 500 068

Walther/Ulm P38 K Series Production 1st Variation (2/1974 – 5/1975)

SN range: 500 101 to 500 500 (400 pistols)

Characteristics: as before, but

- Slide legend on left hand side: "Carl Walther Waffenfabrik Ulm/Do. / P38 K" (with Walther banner and with suffix "K")
- Proof marks: Ulm proof house ("Federal Eagle over N" with "antler" and year stamp "74" resp. "75")
- Safety lever with de-cocking function only – cf. later P4 pistols
- Freestanding barrel held in a sliding barrel carrier; barrel (with its rifling) and barrel carrier are a unit made of one single forged block of steel with a pressed-in liner.
- Barrel carrier guided in slide with one lateral rail on either side (cf. later P5) and guided in frame with one lateral rail on either side – cf. P38 and P4 pistols
- Most pistols with front sight inserted in a dovetail on the slide bridge
- Slide round on top, without a cover – cf. later P4 and P5 pistols

Walther/Ulm P38 K Series Production 2nd Variation (10/1975 – 6/1981)

SN range: 500 501 to 502 595
(2095 pistols)

Characteristics: as before, but

- Most pistols with a fixed, milled front sight on slide bridge; some pistols still with the front sight inserted in a dovetail
- Some pistols with adjustable rear sights
- Slide reinforced – cf. P38/P1 pistols from 11/1967
- Frame reinforced by a hexagonal cross-pin – cf. P38/P1 pistols from 6/1975

Remarks:

Manurhin also assembled short P38 K pistols from parts made by Walther/Ulm. These were marked with "Pistolet P1 K" and sold by Manurhin. There are, however, only very few specimens said to have been made – if any after all.

Engraved P38 K pistols were available on special request.

About 200 P38 K pistols were made in caliber 7.65 mm Parabellum – most of them for the export to Italy (then marked with "CAT 2401" on right hand side of slide).

There were also 9 mm Luger P38 K pistols available with conversion kits for the caliber 7.65 mm Parabellum and vice versa.

The P38 K pistols made for the export to the USA were marked with the Interarms logo on the slide. First it was intended to name these pistols "P38/III", but this intention was rejected later.

Use of Walther P38 K Pistols:

Walther P38 K pistols were issued as service pistols to some police forces (e.g. Munich and Bremerhaven) and the GSG 9 (anti-terror unit of the federal police). They were also used for combat target shooting.

Total production of Walther P38 K pistols: 2,653 pistols

Walther P4 Walther/Ulm and Manurhin

Picture 66: *Walther P4, SN 606 367, series production, caliber 9 mm Luger, produced by Walther/Ulm in 1977, frame and slide reinforced; fixed, no adjustable rear sight*

Production:	3/1975 - 3/1981 (available until 1986)
Calibers:	9 mm Luger (9 x 19 mm) - 8 rounds 7.65 mm Parabellum = .30 Luger (conversion kits only)
Total Length:	7.9 in.
Barrel Length:	4.3 in.
Weight:	28.2 oz.
Rifling:	6 grooves, right-twisted
Operation:	locked blowback breech with firing pin and exposed hammer; pivoting locking block (below the barrel), double action/ single action trigger mechanism

Walther/Ulm P4 transitional model (3/1975 – 8/1975)

SN range: 600 001 to 600 043 (37 pistols, as **SN 600 002 - 600 007** were no P4, but P38 K pistols!)

Characteristics: as Walther/Ulm P38/P1 pistols, but

- Slide legend on left hand side: "Carl Walther Waffenfabrik Ulm/Do. / P38" (with Walther banner)
- Proof marks: Ulm proof house ("Federal Eagle over N" with "antler" and year stamp "75")
- Frame and barrel like P38/P1
- Firing mechanism and shape of slide (round on top) like Walther P4 from series production

Walther/Ulm P4 series production (11/1975 – 3/1981)

SN range: 600 044 to 606 967
(6924 pistols)

Characteristics: as before, but

- Slide legend on left hand side: "Carl Walther Waffenfabrik Ulm/Do. / P4" (with Walther banner)
- Proof marks: Ulm proof house ("Federal Eagle over N" with "antler" and year stamp or year code), resp.
- Acceptance/proof marks on P4s for the German BMI ("BGS") and customs authorities ("BUND"): acceptance stamps of the "Güteprüfdienst des Bundesverteidigungsministeriums" (quality inspectorate of the Ministry of Defense) ("Eagle with Drooping Wings" with identification number of inspector)
- Freestanding barrel held in a sliding barrel carrier; barrel and barrel carrier are a unit made of one single forged block of steel with a pressed-in liner.
- Barrel carrier guided in slide with two lateral rails on either side (cf. later P5) and guided in frame with one lateral rail on either side – cf. P38 pistols
- Slide reinforced – cf. P38/P1 pistols from 11/1967
- Frame reinforced by hexagonal cross-pin – cf. P.38/P1 pistols from 6/1975
- Barrel length: 4.3 in.
- Locking block modified (not identical with the one of P38/P1 pistols)
- "automatic" (shooter independent) safety: After de-cocking the hammer the pistol is automatically in the "safe" mode via a cutout in the hammer: The firing pin is then held down by a spring-loaded push slider. The firing pin is lifted into the reach of the hammer only when the trigger is pulled - similar to the Walther PP Super and the FEG APK resp. the Walther PPK/E – see there.
- Safety lever on left hand side of slide with de-cocking function only (cf. PP-Super), "automatic" safety remains activated after de-cocking;
- Swiveling range of de-cocking lever: 60°
- Slide hold-open lever (on left hand side of frame): activated by the magazine follower, when the magazine is emptied and deactivated manually
- Sights: box-shaped, fixed rear sight and adjustable front sight (on barrel)
- Hammer spur shorter than on P38/P1
- No signal pin (loaded chamber indicator)
- Serrated trigger
- Metal finish: matte-black phosphated finish
- Grip plates: two-piece, made of black, checkered plastic ("Novodur" – a thermoformed ABS plastic made by Bayer) with a little thumb rest (same as on P38/P1 pistols)

Walther/Ulm P4 US Export Model (9/1981)

SN range: 606 968 to 607 419 (452 pistols)

Characteristics: as before, but

- Additional slide legend: "P38 IV"

Remarks:

Walther/Ulm and Manurhin used the same serial range, as Manurhin only marked, blued and assembled the Walther/Ulm made parts – cf. Manurhin Pistolet P1 and Pistolet P1 K.

Some P4 pistols carry the "Sauer & Sohn" logo. These pistols were assembled from Walther made new parts, which the company "Sauer & Sohn" had received from the BMI. In order to get a BMI contract for "S & S" made service pistols this company had to take over used P4 pistols and new spare parts (without serial numbers, but mostly already with BMI and date stamps). About 50 complete "Walther P4" pistols were assembled from these parts. These pistols carry the "S & S" logo on the left hand side of the slide and

Picture 67: *Walther P4, SN P441, caliber 9 mm Luger, assembled by Sauer & Sohn, frame and slide reinforced, BMI and date stamp (3/76) on right hand side of slide*

serial numbers from P401 to about P450.

The last Walther P4 (SN 606 967) was produced in March 1986, but sold in October 1986.

Use of Walther P4 Pistols: Walther P4 pistols were issued as service pistols to the …

- German customs authorities (marked "BUND") and the "Bundesgrenzschutz (BGS – marked "BMI") – proof marks see above. SN range: 600 044 – 605 314 (11/1975 – 5/1976)
- West Berlin police (with Manurhin slide legend and marked with the "Berlin rosette"). SN range: 604 709 – 604 933 (3/1976), taken out of service again in 1979
- "Bereitschaftspolizei" (riot police) of Rheinland-Pfalz and Baden-Württemberg (marked "BMI")
- The Walther P4 was given a NATO supply number (1005-12-169-8552), but was not adopted by the German "Bundeswehr".

Total production of Walther P4 pistols:
7,413 pistols

Walther P5 Walther/Ulm

Picture 68: *Walther P5, SN V1042, experimental pistol, caliber 9 mm Luger, produced by Walther/Ulm in 5/1977, used for drop tests by the "Landespolizeidirektion" Stuttgart*

SN range:	**001 001 to 105 303**
Production:	11/1978 - 1998
Calibers:	9 mm Luger (9 x 19 mm) - 8 rounds 7.65 mm Parabellum = .30 Luger - 8 rounds (1900 pistols only) 9 x 21 mm - 8 rounds (very few pistols only)
Total Length:	7.1 in.
Barrel Length:	3.5 in.
Weight:	28.1 oz.
Rifling:	6 grooves, right-twisted
Operational:	locked blowback breech with firing pin and exposed hammer; pivoting locking block (below the barrel), double action/ single action trigger mechanism

Preliminary remarks:

From February 1976 to October 1977 there were experiments with prototype P5 pistols (V1027 to V1210 – "V" stands for "Versuch" – i.e. experiment or trial). This development was performed with the help of a former SIG pistol designer, Walther Ludwig.

In November 1982 there were even trials with two P5 pistols (V1320 and V1322) in caliber 9 x 18 mm (cf. Walther PP-Super).

The project designation for the later Walther P5 was "P480" first.

The first delivery of P5 pistols (from SN 001 001) in November 1978 went to the police and other government authorities. The commercial market was served only after June 1979.

The development of the Walther P5 pistol was based on the "Pflichtenheft Faustfeuerwaffen" (specifications for handguns), which was issued on 19th June 1975 by the "Technische Kommission" ("TK" – i.e. technical commission) of the

Federal Ministers of the Interior. The relevant trial regulations from the 1st March 1976 demanded functional tests under normal ambient conditions, continuous shooting (10 000 rounds), shooting under difficult conditions and drop tests.

Characteristics:

- Slide legend on left hand side: "Carl Walther Waffenfabrik Ulm/Do. / P5" (with Walther banner); pistols for the commercial market with addition "Made in Germany"; from about SN 100 000 slide legend "P5" only
- Proof marks: Ulm proof house ("Federal Eagle over N" with "antler" and year code), resp.:
- Acceptance/proof marks on P5 pistols for the German BMI ("BGS") and customs authorities ("BUND"): acceptance stamps of the "Güteprüfdienst des Bundesverteidigungsministeriums" (quality inspectorate of the Ministry of Defense) ("Eagle with Drooping Wings" with identification number of inspector)
- Stamp on front of barrel carrier: serial number and caliber (with Walther banner)
- Frame: made of anodized lightweight alloy, reinforced by hexagonal cross-pin – cf. P.38 and P4 pistols
- Slide and minor parts made of blued or phosphated steel
- Freestanding barrel held in a sliding barrel carrier; barrel (with its rifling) and barrel carrier are a unit made of one single forged block of steel without a pressed-in liner.
- Barrel carrier guided in slide with one lateral rail on either side only and no longer guided in frame – cf. P38 pistols
- Slide reinforced – cf. P38/P1 pistols from 11/1967
- Frame reinforced by hexagonal cross-pin – cf. P.38/P1 pistols from 6/1975
- Locking block modified (not identical with the one of P38/P1 pistols)
- Two lateral recoil springs on either side of frame – cf. P38/P1 pistols
- Take down lever on left hand side of frame – cf. P38/P1 and P4 pistols
- Combined slide hold-open and de-cocking lever on left hand side of frame (minor patent, i.e. "Gebrauchsmusterschutz" granted in May 1977)
- Ejection to the upper left hand side
- No signal pin (loaded chamber indicator)
- Sights: adjustable, box-shaped rear sight; fixed front sight on slide bridge
- Trigger with serration and trigger stop
- Safety system: no manual safety, but four "automatic" (shooter independent) safeties:

a) firing pin is locked in idle position and unlocked only if the trigger is pulled and the firing pin is lifted (relevant patent granted in December 1978)
b) hammer has a safety catch
c) hammer has a safety recess at the front, which makes sure that the firing pin cannot be reached by the hammer in lowered, idle position
d) slide closing safety: disconnector makes sure that the trigger bar is disconnected, if the slide is not completely closed

- Bottom magazine release at the heel of the grip
- Grip plates: two-piece, made of black, checkered plastic without a thumb rest; wooden grip plates were available as an option

Walther P5 Walther/Ulm Development and Technical Changes

During the whole production period the Walther P5 pistol underwent a total of 360 (!) minor and major technical changes and improvements by the factory – many of them based on impulses given by representatives from different government authorities. Most of these changes were either retrofitted by police or other armorers resp. by the factory.

The major changes are listed below:

- 07/1979 (from SN 005 701): modification of the hammer and shortening of the firing pin (from 2.63 " to 2.62")
- 11/1979 (from SN 008 401): modification of magazine catch
- 01/1982 (from SN 029 101): additional hole in slide to flush out blueing salts; de-cocking lever reinforced; millings at slide modified (e.g. for the take down lever); rivet on trigger bar reinforced
- 07/1983 (from SN 034 301): spring pin at hexagonal cross-pin omitted, cross-pin from now on caulked only
- 05/1986 (from SN 040 101): magazine catch modified again
- 11/1986 (from SN 044 527): hammer modified again, marked with a four-pointed star for "Form geändert" (design modified) – cf. Walther P1
- 01/1987 (from SN 044 527): take down lever modified, marked with a four-pointed star for "Form geändert" (design modified) and plunger modified accordingly
- 09/1990: "Nachrüstsatz für erhöhte Fallsicherheit" (retrofitting kit for increased drop safety) available, from now on standard in series production
- 10/1990 (from SN 100 730): hammer modified again, marked with a triangle in addition to the four-pointed star.
- 12/1990 (from SN 100 202): milled trigger stop replaced by an adjustable screw, frame reinforced in the grip screw area

Remarks on the drop safety of Walther P5 pistols:

There were some incidents in Holland in 1985 when Dutch police P5 service pistols were dropped and the pistols were fired unintentionally. In a similar accident in Baden-Württemberg on 27th March 1989 a policeman was killed. This led to drop tests by the "Kriminaltechnische Untersuchung" (KTU) in Karlsruhe and similar tests later in the Walther factory.

Together with representatives of the Baden-Württemberg police Walther developed a "Nachrüstsatz zur erhöhen Fallsicherheit" (retrofitting kit for increased drop safety). This kit woks as follows: In the top of the slide the firing pin is held down by a spring-loaded firing pin lock plunger, which makes sure that the firing pin is held in the idle, locked position. For this purpose the firing pin had to be shortened and the upper side of the firing pin had to be modified. The trigger bar and the sear had to be slightly modified as well. From the outside this modification can only be recognized by a small hole in the top of the slide. The relevant patent was filed on 25th April 1990.

As here the trigger bar had to be adjusted together with the sear, this part was no longer interchangeable. So the trigger bars were engraved with the serial number by an electro-pencil.

In April 1990 Walther/Ulm issued a "Umbauanleitung Pistole P5 zur Erhöhung der Fallsicherheit" (retrofitting instructions pistol P5 to increase the drop safety) for the different police forces.

From September 1990 until July 1991 Walther performed this retrofitting in a sort of recall campaign in the factory.

From 1996 to 1998 the Dutch Walther P5 service pistols were retrofitted with this drop safety device as well.

Walther P5 Walther/Ulm Special Models, Contracts and Markings

Picture 69: *Walther P5, SN V1126, experimental pistol, caliber 9 mm Luger, produced by Walther/Ulm in 1977, used for tests by Niedersachsen police, marked "Nds" (for "Niedersachsen")*

Walther P5 pistols for German Police Forces

In the years 1976 and 1977 the Walther P5 was tested together with other competing models by the "Bundesgrenzschutz" academy in cooperation with technicians from various federal state police forces. According to the recommendations of this commission the following federal states introduced the Walther P5 as the standard police service pistol:

a) Rheinland-Pfalz (Rhineland-Palatinate) Police:

- On 13th February 1978: delivery of three P5 sample pistols (serial numbers still prefixed with "V" for "Versuch" – i.e. test)
- From 12th January 1979: delivery of P5 pistols (from SN 002 310)
- In 1999: Rheinland-Pfalz police possessed 10 005 P5, 40 P5 compact and 2 P5 shooting practice barrels

Picture 70: *Walther P1A1, SN V1434, experimental pistol, caliber 9 mm Luger, produced by Walther/Ulm in 6/1988, used for tests by the German "Bundeswehr"*

- The P5 pistol became the standard service pistol of the state police and the riot police (marking for state police: "R-P" (in a circle); marking for riot police: "BMI")

b) Baden-Württemberg Police:

- In 1977: delivery of 25 P5 sample pistols (SN V1100 – 1124)
- From March 1978: delivery of 27 057 P5 pistols (from SN 001 501) and an unknown number of P5 compact pistols for test purposes
- The P5 pistol became the standard service pistol of the state police and the riot police. Marking of state police: "three stacked lions" (in a shield); marking for riot police: "BMI"
- P5 pistols taken out of service again from 2001

c) Niedersachsen (Lower Saxony) Police:

- In 1977: delivery of 12 P5 pistols for tests purposes; five of these pistols were scrapped later, seven were sold on the commercial market (marked "Nds")

Walther P5 Pistols for German "Bundeswehr"

As the Walther P38/P1 was the "Bundeswehr" standard service pistol from 1957 (see there), Walther/Ulm was hoping for a follow-up business: The P5 pistol was modified according to the military specifications and given the service designation "Pistole P1A1" – as the successor to the "Pistole P1". Before already the Walther P5 had been issued a NATO supply number: 1005-12-189-0990 as "Pistole 9 mm P5". This number was, however, not automatically passed on to the pistol P1A1, as the latter was not adopted by the military or any government authority. The Walther in-plant designation for these prototypes was "P5 (P1)".

In June 1980 Walther already experimented with P5 pistols, which had the magazine well, the magazine catch as well as the take down lever of the Walther P38/P1 – see there. This take down lever showed the serrated lever thumb spur angled about 15° to provide a better grip for the thumb.

One of the military specifications was the manual safety. An early P1A1 prototype (SN 044 953) hence has a pivoting safety lever on either side of the slide. Later prototypes – e.g. SN 044 954 already – have a push button as a sliding safety in the top of the slide.

The slide contours of these early prototypes are the ones of the Walther P5 pistols. Only the last prototypes show the slide contours of the P5 compact – i.e. the rounded area at the top is drawn further down on both sides.

A total of 24 pistols P1A1 was made with these specifications and nine of these were shipped to the "Bundeswehr" for tests.

Ultimately the P1A1 was not adopted by the "Bundeswehr" as the new service pistol, however. After tests in Hammelburg, Aachen and Meppen late in 1989 and early in 1990 the "Bundeswehr" finally refused this adoption.

As Walther actually had expected to obtain this contract they had already issued a preliminary instruction manual for the P1A1 pistol.

Characteristics of Walther P1A1 pistol:

- Push button as a sliding safety in the top of the slide
- Slide contours of the P5 compact – i.e. the rounded area at the top is drawn further down on both sides
- Magazine well, magazine catch and take down lever of the Walther P38/P1
- P38/P1 magazine marked "P1"
- Phosphated finish
- SN range: V1422 – V1454 (in discontinuous sequence)

Walther P5 Pistols for Dutch Police ("Rijkspolitie")

A total of 50 000 Walther P5 pistols was procured by the "Intendance Rijkspolitie" ("IRP" – i.e. police administration) in the Netherlands (Holland).

These Dutch P5 pistols were retrofitted with the new drop safety device by Walther/ Ulm between 1996 and 1998.

Characteristics:

- Marking "N" (for Netherlands) on left hand side of slide
- Phosphated finish
- SN range: 050 000 – 100 000 (This SN range is exclusively reserved for the Dutch P5 pistols.)

Walther P5 Pistols for other foreign Police or Armed Forces

The following states introduced the Walther P5 as the standard service pistol for their police or armed forces – often because of their good experiences with the Walther P38/ P1 or PP/PPK pistols:

- Great Britain
- Norway
- Finland
- Sweden
- USA (some police forces only)
- Portugal (army only)
- Last but not least James Bond ("Agent 007") also used the P5 as his service pistol in the films "Octopussy" and "Never Say Never Again" (both produced in 1983). Here the P5 was the successor of his Walther PPK. In modern terms one might speak of a well done "product placement" or even a "surreptitious advertising" here.

Picture 71: *Walther P5-lang, SN V1393, experimental pistol, caliber 9 mm Luger, produced by Walther/Ulm in 1986, with barrel sleeve*

Walther P5-long

This variation was not a success on the commercial market, as it is not really suitable as a target pistol and neither as a defense pistol. So only a few pistols P5-long were produced.

This model was available as a complete pistol, as a conversion kit or as a replacement barrel.

The prototype P5-long pistols still had a barrel sleeve made of alloy.

A total of 75 complete pistols and 21 conversion kits/replacement barrel were manufactured.

Characteristics: as standard P5 pistols, but

- Barrel length: 6 in.
- Barrel with a liner
- SN range: 001 – 095 (96 pistols)

Walther P5 in caliber .22 l.r.

Walther did not really try to come to terms with this variation. Only one conversion kit (SN V1402) was made for tests by the Dutch police.

The gunsmith Johann Uhl in Syrgenstein (near Ulm) manufactured a conversion kit in caliber .22 l.r., however. This system, named "WS-P5", used a split slide and a "Vulkollan" (polyurethane rubber) damper, which was placed behind the right hand side recoil spring. The small bore barrels were provided by Lothar Walther.

Walther P5 Shooting Practice Pistol ("P5-Üb")

Walther developed this model for the police and other government authorities, which had only limited space or financial resources available for shooting practice. Walther designated this variation "P5-Üb" ("Üb" for "Übung" – i.e. practice).

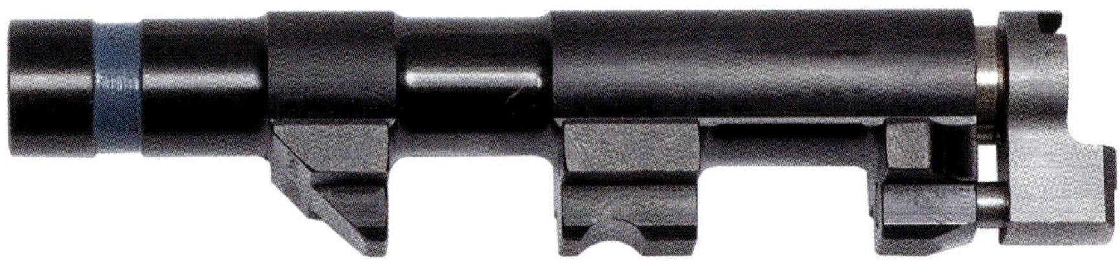

Picture 72: *Walther P5 "PT" barrel, caliber 9 mm PT, produced by Walther/Ulm in 1994*

The practice ammunition for this pistol ("PT" for "Plastik Training") was made of blue plastic and was developed by the companies Geco resp. Dynamit Nobel. The military and police service designation for this type of ammunition is "Übungspatrone 9 mm x 19 DM 58". The idea was to guarantee a practical and even dynamic training without any risks for the environment.

With the first "P5-Üb" pistols the "PT" cartridge was shot using the standard P5 barrel. As then the "PT" ammunition did not provide the necessary momentum for a blowback action, a conversion kit without the standard locking block was developed. Here the barrel unit used a "floating chamber" as a momentum amplifier. Thus the "P5-Üb" could be shot as an unlocked blowback pistol.

For safety reasons practice pistols like the "P5-Üb" are marked with a blue ring near the muzzle. Hence the "PT" ammunition is blue, too. Most, but not all, of the "PT" barrels are longer (4.4 ") than the standard P5 barrels. The chamber of the "PT" barrels differs from the one for the standard 9 mm Luger cartridge. So a 9 mm Luger cartridge does not fit into a 9 mm PT barrel chamber.

The "PT" conversion kit could easily be mounted on a standard frame without any alterations.

Walther made tests with four experimental "PT" barrels (V1211 – V1214) in April 1981 – and again in May 1982 with ten experimental barrels (V1237 – V1246) for the GSG 9 (anti-terror unit of the federal police).

From 1983 to 1989 about 250 "PT" barrels units were produced. From 1987 about 100 complete "P5-ÜB" pistols were made. Most of these practice pistols were issued to the police forces. Only very few were sold on the commercial market.

Some "P5-Üb" pistols were exported – e.g. to the Swiss police in Geneva.

Unfortunately the "P5-Üb" model was not an economic success. This pistol was issued a NATO supply number, however: 1005-12-314-9665. This supply number was used by the police forces, too – be it on the box only.

Walther P5 Laser Shooting Practice Pistol ("Noptel")

The Finnish company Noptel made a shooting practice system ("Noptel ST-15") based on laser technology. The electronic laser unit is inserted into the magazine well. As other practice pistols, too, this model is marked as a practice gun by blue grip plates.

Walther P5 Cutaway Pistols

These pistols were made by Walther/Ulm for teaching and demonstration purposes. They have their own serial number range with a prefixed "S".

SN range: S 152, S 159 S 182, S 347 as well as **S 416 - S 418** are known so far

Picture 73: *Walther P5, SN S 347, cutaway pistol, caliber 9 mm Luger, produced by Walther/Ulm*

Walther P5 Jubilee Edition ("100 Jahre Walther")

This limited edition was issued in 1986 to celebrate the 100 years anniversary of the Walther company.

Characteristics:

- Slide legend on left hand side of slide: "100 Jahre Walther / 1886 – 1986" (the word "Walther" with gold inlay)
- Trigger, safety and hammer gold-plated
- Metal finish: high-gloss polished finish
- Carved wooden grip plates
- Presentation case

Walther P5 engraved

Here the same is true as for the PP/PPK and P.38 models: There was and is no factory engraving department in Ulm at all. So the much quoted "factory engravings" simply do not exist. There were engraved pistols sold by Walther/Ulm, but these engravings were all made by subcontractors – first by Josef Aichholzer, later by Dieter Stübler (both located in Ulm).

Walther P5 with "Steyr Mannlicher" slide legend

In February 1990 Walther delivered two P5 pistols (SN 050 201 and 050 203) to the company Steyr Mannlicher in Austria. They carry the slide legend "Steyr Mannlicher GesmbH" on the left hand side of the slide. These two pistols were samples for an intended "joint venture" of Walther/Ulm and Steyr Mannlicher to arm the Austrian "Gendarmerie" with Walther P5 pistols. In January 1992 these two pistols were sold to Steyr Mannlicher employees.

Total production of Walther P5 pistols:
104,303 pistols

Walther P5 Compact Walther/Ulm

Picture 74: *Walther P5 compact, SN 150 004, experimental pistol, caliber 9 mm Luger, produced by Walther/Ulm in 7/1987, 4th pistols of series production*

SN range:	**150 001 to 155 569**
Production:	7/1987 - 1996
Calibers:	9 mm Luger (9 x 19 mm) - 8 rounds 9 x 21 mm - 8 rounds (very few pistols only)
Total Length:	6.7 in.
Barrel Length:	3.1 in.
Weight:	26.4 oz.
Rifling:	6 grooves, right-twisted
Operation:	locked blowback breech with firing pin and exposed hammer; pivoting locking block (below the barrel), double action/ single action trigger mechanism

Characteristics: as Walther P5, but

- Slide legend on left hand side: "P5 compact" (with Walther banner); pistols for the commercial market with addition "Made in Germany"; from about SN 100 000 slide legend "P5" only
- Proof marks: Ulm proof house ("Federal Eagle over N" with "antler" and year code), resp.:
- Acceptance/proof marks on P5 pistols for the German police or other government authorities: acceptance stamps of the "Güteprüfdienst des Bundesverteidigungsministeriums" (quality inspectorate of the Ministry of Defense) ("Eagle with Drooping Wings" with identification number of inspector)
- Bottom magazine release at the heel of the grip (mostly for police and other government authorities) or upper magazine release (convertible for left-handed and right-handed shooters)
- Main spring in a housing in backstrap
- Front and back of grip serrated
- Grip plates: two-piece, made of black, checkered plastic without a thumb rest

(for authorities) or made of smooth resp. checkered wood (for the commercial market)
- Metal finish: phosphated (for authorities) or matte black blued (for the commercial market)
- No barrel liner
- From 9/1990: "Nachrüstsatz zur erhöhen Fallsicherheit" (retrofitting kit for increased drop safety) – cf. Walther P5

Remarks:

From September 1983 a pre-series of nine experimental pistols was made (V1171 – V1396, serial numbers in discontinuous sequence)

In February 1990 Walther delivered two P5 compact pistols (SN 151 394 and 151 395) as well as to two P5 pistols (see there) to the company Steyr Mannlicher in Austria. They carry the slide legend "Steyr Mannlicher GesmbH" on the left hand side of the slide. These pistols were samples, too, for an intended "joint venture" of Walther/Ulm and Steyr Mannlicher to arm the Austrian "Gendarmerie" with Walther pistols. In January 1992 these two pistols were sold to Steyr Mannlicher employees.

Use of Walther P5 compact Pistols:

These pistols were and are often used for combat target shooting practice as their smooth outside contours allow a quick drawing. Moreover the operating levers are conveniently close together on the left hand side of the pistol.

Police forces and other authorities, which had introduced the Walther P5 as a service pistol, often later procured P5 compact pistols, too, for officers with small hands.

Walther P5 compact for British Army and Police Forces ("L102A1")

From 1989 the Walther PP and PPK pistols as "Personal Protection Weapons" (PPW – see there) of the British army and police were gradually taken out of service. The SIG pistol P230 was introduced as pistol

Picture 75: *Walther P5 compact ("L102A1"), SN 172 673, caliber 9 mm Luger, produced by Walther/Ulm in 1992, British "Personal Protection Weapon"*

"L109A1" and the Walther P5 compact as pistol "L102A1". The letter "L" here stands for "landforces". The Walther P5 compact was given the NATO supply number B1/1005-99-978-4952.

In April 1989 Walther delivered a sample pistol (SN V1491) to Great Britain. Although the FN HP 9 mm pistol ("L9A1") was still the standard service pistol, the British armed forces procured a total of 3500 Walther P5 compact pistols. These were issued as "Personal Protection Weapons" (PPWs) to members of the "Ulster Defence Regiment" (from 1992 renamed "Royal Irish Regiment") and other British troops in Northern Ireland. Anti-terror units like the Special Air Service (SAS) and some pilots of the British air force were equipped with Walther P5 compact pistols as well.

The British even issued special belt holsters for the concealed carrying of the P5 compact pistols with the NATO supply numbers 1095-99-978-4915 (for right-handed shooters) and 1095-99-978-4916 (for left-handed shooters).

Characteristics: as standard P5 compact pistols, but

- Slide legend on left hand side: "L102A1" (with Walther banner)
- Upper magazine release
- Phosphated finish

Total production of Walther P5 compact pistols: 5,569 pistols

Walther P88 Walther/Ulm

Picture 76: *Walther P88, SN 001 003, caliber 9 mm Luger, produced by Walther/Ulm in 1/1987, 3rd pistol of series production*

SN range:	**001 001 to 010 894**
Production:	1984 resp. 1987 - 1992
Calibers:	9 mm Luger (9 x 19 mm) - 15 rounds
Total Length:	7.5 in.
Barrel Length:	3.9 in.
Weight:	31.5 oz.
Rifling:	6 grooves, right-twisted
Operation:	locked blowback breech with firing pin and exposed hammer; tilting barrel (modified Browning system with a fixed, open cam), double action/ single action trigger mechanism

Preliminary remarks:

For Walther the P88 pistol represents an important step in their weapon technology. On the one hand this pistol model is the last full metal type, on the other hand it is the first Walther pistol with the Browning tilting barrel system. The upper part of the chamber serves as a locking block, which is supported in the slide against the front of the ejection port.

The development of the P88 pistol was essentially influenced by the former SIG pistol designer Walter Ludwig – cf. SIG P220 pistol.

Walther P88 experimental and pre-series

The first experimental pistols were manufactured from 1980 with the project designation "P485" (SNs in discontinuous sequence V1170 – V1195). In October 1981 a further eight experimental pistols followed (SNs in discontinuous sequence: V1223 – V1408). In November 1981 more experimental pistols were made: V1536 – V1549 and 88/1379 – 88/1389 (SNs not "V" prefixed!). In December 1981 a pre-series followed: SN 881 200 – 881 390.

Amongst others some of these pistols were made for tests by the US army in Aberdden, as the US was searching for a successor to the Colt Government 1911A1 service pistol as pistol "M9". This "Joint Services Small Arms Program" (JSSAP XM9) began in 1981 and ended in 1985.

Characteristics: differing from later series production P88 pistols

- Frame and slide not slimmed at the front and about 20 mm longer
- Backstrap straighter
- Metal finish matte blued
- Some pistols with wooden grip plates, held by one grip screw only
- Trigger bar narrower

Walther P88 series production

The series production of P88 pistols started in January 1987 with SN 001 001.

Characteristics:

- Slide legend on left hand side: "P88 Made in Germany" (with Walther banner); pistols for the commercial market with addition "Made in Germany"; from about SN 100 000 slide legend "P88" only
- Proof marks: Ulm proof house ("Federal Eagle over N" with "antler" and year code)
- Frame made of black anodized lightweight alloy with steel reinforcements
- Slide and minor parts made of matte black blued steel, sides of slide polished in the slide legend area
- Matte nickel-plated finish available as an option
- Tilting barrel (modified Browning system with a fixed, open cam)
- Recoil spring below barrel on a guide rod
- Take down lever on left hand side of frame
- Ambidextrous, combined slide hold-open and de-cocking lever on left hand side of frame (minor patent, i.e. "Gebrauchsmusterschutz" granted in May 1977)
- Ejection to the top
- No signal pin (loaded chamber indicator)

- Sights: adjustable, box-shaped rear sight; fixed front sight on slide (up to 12/1987 inserted in a dovetail; from 12/1987 as a fixed, square post sight); "3-dot" paint marking
- Trigger with an adjustable trigger stop
- Safety system: no manual safety, but four "automatic" (shooter independent) safeties:

a) firing pin is locked in idle position and unlocked only if the trigger is pulled and the firing pin is lifted (relevant patent granted in December 1978)
b) hammer has a safety catch
c) hammer has a safety recess at the front, which makes sure that the firing pin cannot be reached by the hammer in lowered, idle position
d) slide closing safety: disconnector makes sure that the trigger bar is disconnected, if the slide is not completely closed

- Ambidextrous, upper magazine release behind the trigger
- Grip plates: two-piece, made of black, checkered plastic without a thumb rest, two grip screws
- Two row magazine, capacity: 15 rounds

Use of Walther P88 Pistols:

With their new P88 pistol Walther had hoped to obtain the contract for the new US army service pistol "M9". As the P88 has no manual safety the US decided against this pistol model and adopted the Beretta Mod. 92F.

In 1991 the police of the Netherlands (Holland) received two P88 pistols (V1536 and V1537) to be tested as a successor to the Dutch Walther P5 service pistols. But they did not adopt this pistol model either.

The P88 pistol was only used as a service pistol by some special forces of the German "Bundesgrenzschutz" (BGS).

Due to the relatively high price, the P88 pistols could only gain a niche position on the commercial market, despite high quality standards.

Total production of Walther P88 pistols: 9,894 pistols

Walther P88 Compact Walther/Ulm

SN range:	**100 001 to 107 344**
Production:	2/1992 - 2000
Calibers:	9 mm Luger (9 x 19 mm) - 14 rounds
Total Length:	7.1 in.
Barrel Length:	3.8 in.
Weight:	29.0 oz.
Rifling:	6 grooves, right-twisted
Operation:	locked blowback breech with firing pin and exposed hammer; tilting barrel (modified Browning system with a fixed, open cam), double action/ single action trigger mechanism

Preliminary remarks:

From February 1987 tests with 13 experimental P88 compact pistols (V1411 – V 1560 and V1571) were made, before in February 1992 the series production began with SN 100 001.

Picture 77: *Walther P88 compact, SN 105 129, caliber 9 mm Luger, produced by Walther/Ulm in 1996*

Characteristics: as Walther P88, but

- Compacter dimensions and smoother outer surfaces
- Recoil spring made of twisted spring wire
- No ambidextrous, combined slide hold-open and de-cocking lever on left hand side of frame, but only slide hold-open lever on left hand side of frame
- No "automatic" (shooter independent) safety, which locks the firing pin in idle position, but ambidextrous, manual safety in the rear of the slide. This safety locks the firing pin when in position "safe" and also has a de-cocking function (similar to the P38/P1 pistols)
- Lanyard loop at the heel of the grip

Remarks:

Like Walther P88 the P88 compact came a bit late onto the market and thus had to compete with pistols, with polymer frames. These could be produced more cheaply. Both the P88 and the P88 compact pistols were high-quality and precise weapons, but simply too expensive in production.

Although the P88 compact is equipped with a manual safety to meet the military and police specifications, this model was not adopted by any armed or police force.

Walther P88 compact Special Variations

Walther "P88A1"

Late in 1989, i.e. clearly before the start of the series production, Walther produced a pre-series of ten P88 compact pistols marked "P88A1" (V1496 – V1502). These pistols were made for the "Bundeswehr" tests in Hammelburg, Aachen and Meppen late in 1989 and early in 1990. Strangely enough then this "P88A1" competed with another Walther pistol: the "P1A1".

Walther issued a preliminary instruction manual for the P1A1 pistol, too.

As Walther actually had expected to obtain this contract they had already issued a preliminary instruction manual for the P1A1 pistol.

Characteristics: as standard P88 compact, but

- Slide legend on left hand side: "P88A1 / 9 mm x 19 / 12/89" (with Walther banner)
- Frame marked "Bw" on left hand side
- Frame slightly slimmed at the front
- Extractor serves as loaded chamber indicator
- Semi-circular milled-out slide serrations in front of the safety lever on left hand side of slide
- Phosphated finish
- Matte black plastic grip plates – either stippled or checkered

Walther "P88B" ("Behördenmodell")

Only three experimental pistols (V1616 – V 1618) were made of this variation.

Characteristics: as standard P88 compact, but

- Slide legend on left hand side: "P88 compact B" (with Walther banner); the letter "B" stands for "Behörden" (government authorities)
- Ambidextrous safety lever on either side

Walther "PC Police"

Only one experimental pistol (V1619) of this variation was made in November 1995. Although Walther advertised this model in a brochure, it was never produced in series production.

Characteristics: as standard "P88 compact B", but

- Slide legend on left hand side: "PC Police" (with Walther banner); the letter "B" stands for "Behörden" (government authorities)

Total production of Walther P88 compact pistols: 7,344 pistols

Picture 78: *Walther P88 compact ("P88A1"), SN V1497, caliber 9 mm Luger, produced by Walther/Ulm in 1989, used for tests by German "Bundeswehr"*

Picture 79: *Walther P88 compact ("PC Police"), SN V1619, caliber 9 mm Luger, produced by Walther/Ulm in 1994*

Walther PP90 Walther/Ulm

SN range:	**V1522 - V1523 and V1531 - V1535** (7 pistols)
Production:	1990 only
Calibers:	9 mm Luger (9 x 19 mm) - 14 rounds
Total Length:	7.5 in.
Barrel Length:	3.6 in.
Weight:	37.2 oz.
Rifling:	6 grooves, right-twisted
Operation:	locked blowback breech with firing pin and exposed hammer; long, pivoting locking latch below the barrel, double action/ single action trigger mechanism

Preliminary remarks:

Sam Cummings, the US importer of Walther pistols, suggested to develop an "enlarged PP pistol" with a bigger magazine capacity for the US market.

So the model designation of the PP90 is programmatic in itself. The outlines, too, should suggest a further technical development of the Walther PP pistol – in the more powerful caliber 9 mm Luger.

In the past, two attempts to design an unlocked Walther pistol in this caliber had both failed, however – cf. Mod. 6 and "Militär-Pistole". So the new design had to be a locked blowback pistol.

But somehow the model PP90 does not fit into the Walther pistol model range at all. Neither does it meet the demands of the late 1990s weapons market. This pistol is made of steel again and it is a step backwards – away from the Browning tilting barrel systems again.

When designing the Walther PP90 pistol Walter Ludwig borrowed some features from the Italian Sosso pistol made in 1934. This

Picture 80: *Walther PP90, SN V1534, caliber 9 mm Luger, produced by Walther/Ulm in 1990*

model had a long locking latch below the barrel, too.

Characteristics:

- Slide legend on left hand side: "PP90" (with Walther banner)
- Proof marks: Ulm proof house ("Federal Eagle over N" with "antler" and year code
- Frame, slide and minor parts made of blued steel
- Slide with 25 serrations on either side (in front of the safety levers)
- Barrel unit consisting of a barrel, a barrel carrier, a moveable ring element and a long, pivoting locking latch below the barrel unit
- Barrel serves as a guide for the recoil spring
- Slide hold-open lever (in left hand side of frame) is activated by the magazine follower, when the magazine is emptied – cf. Walther PP/PPK, but is deactivated manually
- Trigger module with a forked trigger bar (module removable from the frame)
- Ambidextrous take down latch on frame (in front of the trigger)
- Ejection to the right hand side
- No signal pin (loaded chamber indicator)
- Sights: rear sight inserted in a dovetail; fixed front sight on slide; "3 Dot" color marking
- Trigger with a fixed trigger stop
- Safety system:

a) firing pin is locked in idle position and unlocked only if the trigger is pulled and the firing pin is lifted by the sear
b) hammer has a safety catch
c) trigger bar is disengaged from sear by a spring-loaded plunger when in idle position
d) ambidextrous manual safety levers (similar to P88 compact), without de-cocking function though

- Ambidextrous, upper magazine release behind the trigger

- Grip plates: two-piece, made of smooth wood, without a thumb rest, two grip screws
- Two row magazine, capacity: 15 rounds (from P88 compact)

Remarks:

Due to its elaborate design this pistol model was relatively expensive in production. After endurance firing tests there were even fissures on the locking latch.

To make a bad story worse, Walther/Ulm was in a difficult financial situation in 1990 and the US market became more and more problematic. So these and other reasons prevented a series production of the Walther PP90 pistol.

You will find a detailed and illustrated description of this pistol model in: Dieter H. Marschall, "Shooting Star", in: Deutsches Waffen-Journal 6/2002.

Total production of Walther PP90 pistols: 7 pistols

Walther P99 Walther/Ulm

Picture 81: *Walther P99, SN V1639, caliber 9 mm Luger, experimental pistol, produced by Walther/Ulm in 8/1996, early frame with "M" marking on left hand side, "first generation" P99 pistol*

SN ranges:

from 001 001 (11/1996) in 9 mm Luger
from 400 001 (02/1998) in .40 S&W
from 100 001 (02/1998) in 9 x 21 mm

Production: 11/1996 - to date

Calibers: 9 mm Luger (9 x 19 mm) - from 1996
9 x 21 mm (especially for export to Italy) - from 1998

	10 x 21 mm (= .40 S&W) - from 1998
	.45 ACP (only S&W production) - from 3/2003
Total Length:	7.1 in. (S&W: 7.2 in)
Height:	5.3 in.
Barrel Length:	4.0 in. (S&W: 4.1 in)
Weight:	24.2 oz. (S&W: 25.9 oz.)
Rifling:	6 grooves, right-twisted
Operation:	locked blowback breech, hammerless, striker-fired; short recoil, tilting barrel (modified Browning system); double action/ single action trigger mechanism

Characteristics:

- Slide legend on left hand side: "P99" (with Walther banner), later P99 pistols with additional identification letters for the different variations
- Proof marks: Ulm proof house ("Federal Eagle over N" with "antler" and year code)
- Frame made of polymer plastic (reinforced by glass or carbon fibers) with embedded steel guide rails for the slide and a burled surface
- Accessory rail (later standardized Weaver resp. Picatinny rail) at the front of the frame (for e.g. a laser sight module)
- Interchangeable backstraps (in three sizes)
- Slide and minor parts made of blued or tenifer treated steel
- Slide serrations on both sides at the rear of the slide (with some variations at the front, too), number of serrations varies depending on the variation
- Slides for P99 pistols in caliber .40 S&W are 4 mm longer and have no slanted contour at the front
- Tilting barrel (modified Browning system with a fixed, open cam), i.e. the upper part of the chamber serves as a locking block, which is supported in the slide against the front of the ejection port
- Recoil spring made of flat spring wire, below barrel on a guide rod
- Slide stop lever first on left hand side of frame only, later ambidextrous
- Magazine release levers on both sides of trigger guard
- De-cocking button in top of slide: first on left hand side only, later ambidextrous
- Ambidextrous take down catch in the frame (in front of the trigger)
- Ejection to the top
- Loaded chamber indicator: When a cartridge is in the chamber, the rear of the extended extractor is withdrawn into the slide; a red dot then becomes visible; the hook of the extractor then protrudes from the slide and becomes perceptible
- Cocking indicator (except for P99 DAO): red-dotted, rear end of striker serves as cocking indicator (cf. Walther Mod. 9)
- High contrast sights: adjustable rear sight with phosphorescent paint markings (first U-shaped, later "3-dot" system) and interchangeable front sights (first made of plastic, later made of steel); Tritium night sights available for law enforcement authorities only
- Trigger module removable from the frame
- Trigger system: first DA/SA trigger (so-called "TA" trigger for "traditioneller Abzug", i.e. traditional trigger), later other trigger variations – see below
- Safety system: no manual safety, but four "automatic" (shooter independent) safeties:

a) striker is locked in idle position by a spring-loaded bolt and unlocked only if the trigger is pulled; this serves as a drop safety as well

b) trigger safety blocks the trigger in the frame, if no pressure is exerted on the trigger; some pistols with an additional middle trigger safety (integrated release lever in trigger)

c) de-cocking safety: de-cocking lever makes sure that the pistol can be de-cocked safely

d) additional drop safety (with P99 DAO only): makes use of the inertia momentum blocking the striker

- Two row magazine, capacity: 10, 12, 15 resp. 16 rounds

Remarks:

The Walther P99 pistols was first designated "HLP" (for "hammerless pistol").

This pistol model brought with it a radical change in the Walther pistol model range, as it was the first one with a plastic frame. In 1970 the German company Heckler & Koch/ Oberndorf had used this technology first with their pistol model VP70 – without great success. The real breakthrough of this new production method came with the Glock pistol Mod. 17 in 1982.

Pistols with polymer frames are often ridiculed as "Tupperware pistols", but the data speak for themselves: In 2000 the production of an (empty) P99 frame cost only DM 8.00 (about 4.00 EUR), whereas a P5 frame cost about DM 175.00. The production of a complete P5 pistol took about eight hours, whereas the P99 takes only 2.1 hours. Moreover, the polymer frames are much lighter and more durable than metal frames.

Of course it made no sense for Walther to produce these frames themselves. Thus an intensified "outsourcing" became necessary. Whereas the degree of vertical production integration (i.e. the percentage of parts made in-house) with the Walther P5 was about 60 percent, it is now only 30 percent with the P99 pistols. This means, as well, that more quality control measures are necessary for incoming pistol parts.

Only pistols produced with this new technology can be competitive on today's market. In the tendering procedures of most police forces e.g. the award of a contract depends more on the price (about 45 %) and less on the quality (only about 15 %!). The performance and technical specifications of the government authorities of course contain some exclusion criteria.

The Walther P99 therefore was designed according to the "Technische Richtlinien Pistole" (technical specifications for pistols) of the German police forces.

The new Walther pistol designer developing the "HLP" resp. P99 pistol was Horst Wesp from Styria (Steiermark) in Austria. His work began in September with a prototype (V1615). In February 1995 followed a small pre-series of experimental pistols (V1626 – V 1628), followed again by a second pre-series in November 1995 (V1631 – V1645).

From June 1996 Sam Cummings, the US importer from Interarms, received the first shipment of sample pistols (V1637 – V1654). In the summer and autumn of 1996 a number of sample pistols went to resellers in Germany and abroad as well as to government authorities (V1655 – V1856).

In November 1996, shortly before the market launch, a misfortune happened in production with pistols SN V1861 – V1932: The serrations at the rear end of the slide were milled in the wrong angle (inclined rearwards). Some of these slides were reworked by milling the serrations in the right angle – ending up with a diamond-shaped pattern. These pistols were sold more cheaply as second choice items.

By the end of November 1996 the Walther P99 in caliber 9 mm Luger had reached its market readiness. Experiments and tests with P99 pistols in other calibers and variations followed:

P99 .40 S&W:	12/1996 (V1935 - V1938) and (V1948 - V1955)
P99 9 x 21 mm:	02/1997 (from V1939)
P99 "La Chasse":	10/1997 (V1956 - V1985)
P99 DAO:	02/1998 (from V1966)

Picture 83: *Walther P99 QA, SN FAD 5833, caliber 9 mm Luger third generation" P99 pistol*

another 3 mm and the pistol is fired; all shots made with pre-cocked striker, i.e. with the same trigger weight and path from the first to the last shot

- Small de-cocking button only, as it is only needed to de-cock the pistol before disassembly without pulling the trigger

P99 Surface Finishes

Remarks:

There are different combinations of frame colors and metal finishes possible with P99 pistols. The P99 compact comes in black/ black only.

P99 frames

- Standard color: black
- "military" variation in NATO olive green or sandy-colored available

P99 slides

- Tenifer treated: black surface by nitrocarburizing in a salt melt (standard with P99 pistols from 1996)
- QPQ ("Quench Polish Quench"): special silver-colored surface treatment (not coating!) by intermittent polishing during tenifer treatment; other operating elements black, i.e. bi-color appearance
- Titanium coated: silver-colored coating (available from 2002)

Walther P990 (from 1999)

Remarks:

In 2003 Walther/Ulm delivered 20 sample pistols type P990 DAO in 9 mm Luger and 15 pistols in .40 S&W to Smith & Wesson. These pistols carried the slide legend "Smith & Wesson, Springfield, MA" on the right hand side of the slide.

Standard P990 pistols with the characteristics as follows:

Characteristics: as standard P99 pistols, but

- Available in calibres 9 mm Luger or .40 S&W
- Trigger module: either DAO or QA – see

Picture 84: *Walther P99 AS, SN 412 631, caliber .40 S&W, produced by Walther/Ulm in 2006*

above
- Short ambidextrous magazine release levers
- Short slide stop lever on left hand side of frame
- No de-cocking button (with DAO pistols) or only small de-cocking button (with QA pistols) in top of slide
- Sights: adjustable rear sight with phosphorescent paint markings (U-shaped), interchangeable front sights
- Some pistols for USA with magazine safety
- "P990 Military" available with NATO olive green frame

Walther P99 QA "Quick Action" (from 2000)

Remarks:

In 1999 Walther/Ulm introduced this P99 variation as P99 QSA (for "Quick Safe Action"). On 1st July 1999 two pistols (SN 023 060 and 023 061) in this variation (still with the big de-cocking button) were shipped to the "Polizeibeschaffungsamt" (police procurement authority) in Hildesheim/Lower Saxony for tests.

Standard P99 QA pistols from 2000 with the characteristics as below (in 9 mm Luger from SN 500 001 and in .40 S&W from SN 480 001).

Characteristics: as standard P99 pistols, but
- Available in calibres 9 mm Luger (from SN 500 001) or .40 S&W (from SN 480 001)
- Trigger module: Quick Action (QA) – see above
- Series production P99 QA: small, shrouded de-cocking button only, as de-cocking is only necessary before disassembly; to make sure that this button is not pressed accidentally, it can only be operated using a pointed pin
- First only one short slide stop lever on left hand side of frame, later extended ambidextrous levers on both sides
- "P99 QA Military" available with NATO olive green or sandy-colored frame

- Available for authorities in 9 mm Luger with steel sights and ambidextrous slide stop levers
- From 2004: available in variation "P99 QA Titanium" with titanium-coated slides

Walther P99 AS "Anti Stress" (from 2001)

Characteristics: as standard P99 pistols, but

- Available in calibers 9 mm Luger or .40 S&W
- Trigger module: Anti Stress (AS) – see above
- Wide de-cocking button in top of slide
- First only one short slide stop lever on left hand side of frame, later extended ambidextrous levers on both sides
- First only short, ambidextrous magazine release levers; later extended, ambidextrous levers
- Available for authorities with extended, ambidextrous slide stop levers and extended, ambidextrous magazine release levers
- 2003: delivery of ten P99 AS sample pistols to the Catalan police in Spain (marked: "Policia Mossos d'Esquadra", i.e. a militarized police corps)
- 2008: delivery of P99 AS pistols to the South-African police forces ("SAP") with additional ambidextrous safety levers in the frame (below the slide stop levers)

Walther P99 TA "Traditioneller Abzug" (from 2001/2002)

Remarks:

Only 50 P99 TA pistols were manufactured to be tested by the Baden-Württemberg police. The government of this federal state decided however not to introduce this model as their new police service pistol.

Characteristics: as standard P99 pistols, but

- Available in caliber 9 mm Luger
- Trigger module: named "TA", but actually it is an "AS" trigger system with DA/SA mode – see above
- Wide ambidextrous de-cocking button in top of slide
- Extended ambidextrous slide stop levers on both sides
- Short, ambidextrous magazine release levers
- Frame modified at the front for an easier holstering
- Some pistols with additional slide serrations at the front
- Rear sights with U-shaped phosphorescent paint markings

Walther P99 DAO "Double Action Only" (from 2004)

Remarks:

This model is basically a further development of the P990. On the "Internationale Waffenausstellung" (2002 in Nuremberg) this variation was still presented as P99 DA.

In 2004 ten P99 DAO pistols in caliber 9 mm Luger were made as sample pistols for the Nordrhein-Westfalen police.

In 2005 the "Bundeszollverwaltung" (federal customs administration) received three P99 DAO pistols in caliber 9 mm Luger for tests (marked "BZV" on right hand side of slide).

Characteristics: as standard P99 pistols, but

- Available in calibers 9 mm Luger or .40 S&W
- Trigger module: "DAO" – see above
- No de-cocking button in top of slide
- Extended ambidextrous slide stop levers on both sides
- First short and later extended, ambidextrous magazine release levers

Picture 85: *Walther P99 D, caliber 9 mm Luger, sample pistol for German authorities, "third generation P99", photo: Carl Walther/Ulm*

- Some pistols with an additional small "trigger in the trigger"
- Sights: optional steel sights ("3-dot" system) or Tritium night sights (available for authorities only)
- "P99 DAO Military" available with NATO olive green or sandy-colored frame
- P99 DAO pistols are only produced for authorities on special request.

Walther P99 Q (from 2008)

Remarks:

This P99 Q model is basically a further development of the P99 QA and represents the fourth generation of the P99 model range.

Characteristics: as standard P99 pistols, but

- To date available in caliber 9 mm Luger and for German or foreign authorities only
- Developed according to the "Technische Richtlinien Pistole" (technical specifications for pistols) of the German police forces, i.e. trigger with constant trigger weight (30 Newtons) and pull (12 mm) from the first to the last shot;
- Pre-tensioned striker spring and partially cocked trigger ("Quick Defense")
- Double action to make sure that, if a cartridge fails, the trigger can be pulled again without having to retract the slide by hand
- New design of frame: Picatinny rail (according to MIL-STD-1932), "Hi-Grip" surface with a sanded texture, built-in service chip, lanyard loop at the heel of the grip; model designation on both sides of the frame
- New design of slide: top more rounded, no slanting surfaces or corrugations (so-called "softline design"), seven coarse serrations at the front and eight at the rear of the slide, no de-cocking button in top of slide
- Extended ambidextrous slide stop levers on both sides
- Extended, ambidextrous magazine release levers
- High contrast sights: adjustable rear sight

with phosphorescent paint markings ("3-dot" system) and interchangeable front sights
- Magazine safety
- Magazines: 15, 16 or 20 rounds
- In January 2008 the P99 Q was certified according to the "Technische Richtlinien Pistole" (technical specifications for pistols) of the German police forces

Walther P99 D (from 2008)

Remarks:

This P99 D model is basically a further development of the P99 DAO and like the P99 Q this pistol represents the fourth generation of the P99 model range. The P99 D was developed for government and other law enforcement authorities. There is, however, a trend perceptible within these authorities to prefer the QA and no longer the DAO variation. So the P99 D pistol model was not produced in series production.

Characteristics: as standard P99 Q pistols, but

- Available in caliber 9 mm Luger
- Trigger module: "DAO" – see above
- Trigger interchangeable (two sizes)
- Trigger guard enlarged

Walther P99 Special Models

Walther P99 SD (silencer)

Characteristics:

- As standard P99 pistols, but
- No complete pistols made, but only longer (4.4") replacement barrels with a left-hand thread at the front
- Legend on right hand side of chamber: "Walther 9 mm x 19 SD" (with Walther banner)
- SD-barrels for the USA with additional legend "Earl's Repair SVC" on right hand side of chamber (Earl J. Sheehan Jr. is an importer of Walther weapons to the USA)
- Screw-on ring as a thread protector
- Silencers provided by Brügger & Thomet in Switzerland (now named B&T AG)

Walther P99 S ("S" for Switzerland?)

Characteristics:

- As standard P99 pistols, but
- Additional slide serrations at the front of the slide (similar to the later PPQ) – see there
- Only 15 pistols made for a Swiss police force
- SN range: 043 899 – 043 914

Picture 86: *Walther P99 SD, SN 050 307, caliber 9 mm Luger, produced by Walther/Ulm in 2002*

Picture 87: *Walther P99 "La Chasse", SN V1985, caliber 9 mm Luger, produced by Walther/Ulm in 1998*

Walther P99 "La Chasse" (1998 – 2000)

Remarks:

The P99 "La Chasse" (in English: "The Hunt") was designed especially for the needs of hunters in cooperation with Hans-Joachim Duderstaedt. In October 1997 Walther made the first experimental P99 "la Chasse" pistols (V1956 – V1965).

Characteristics:

- As standard P99 pistols, but
- Available in the calibers 9 mm Luger, .40 S&W or 9 x 21 mm (the latter is very rare!)
- Frame and slide olive green
- Trigger module: "Anti-Stress" (AS) – see above
- Standard sights: steel rear sights ("3-dot" system) and interchangeable front sights; optional sights: so-called "Fluchtvisier" (i.e. quick target acquisition sights) with embedded red fiberglass in front and rear sights
- Slide with optional hand or laser engravings (scrolls or oak leaves) and/or the owner's name engraved on the left hand side
- Backstrap made of stippled walnut wood, interchangeable backstraps made of plastic
- Olive green lanyard
- Olive green luxury carrying case with accessories
- Total production: 439 PP99 "La Chasse" pistols (128 of these for export to the USA)
- SN range: 002 001 – 002 439

Walther P99 Special Edition "James Bond" (1998)

Remarks:

In 1997 the co-owner of Walther/Ulm, Wulf-Heinz Pflaumer, presented the James Bond actor, Pierce Brosnan, his personal P99 service pistol for the shooting of the film "Tomorrow Never Dies". After all James Bond was perhaps not faithful to his playmates, but from 1958 to 1997 he was relatively faithful to his Walther service pistols – with some short "secret escapades". From 1997 onwards he used his new Walther P99 but also his still his old Walther PPK pistol. Interestingly enough James Bond intermittently posed with his long-barreled Walther LP53 air (!) pistol – to the scare of his adversaries?

Characteristics: as standard P99 pistols, but

- Available in calibers 9 mm Luger (from SN 007 001) and .40 S&W (from SN 007 501) – all serial numbers starting with "007 …"
- Limited edition: 1000 pistols
- Slide legend on left hand side: "Her Majesty's Secret Intelligence Service" (in a banner-shaped emblem, which is not the official emblem of the British MI6, however) and "One of 1000" (both as laser engravings), Walther banner moved to the rear of the slide
- Luxury case with accessories

Walther P99 Jubilee Edition "Jahr 2000" (1999)

Characteristics:

- As standard P99 pistols, but
- Available in calibers 9 mm Luger (from SN 2000 1001) and .40 S&W (from SN 2000 0001) – all serial numbers starting with "2000 …"
- Limited edition: 2000 pistols
- Slide legend on left hand side: "P99 Commemorative for the Year 2000" and "2000 [plus SN] of 2000" (both as laser engravings)
- Luxury case with accessories

Walther P99 Jubilee Edition "120 Jahre Walther" (2006)

Characteristics:

- As standard P99 pistols, but
- Slide legend on left hand side: "120 Jahre 1886 – 2006 / P99" (inscriptions with golden inlays)
- Engravings (scroll pattern)
- Luxury case with accessories

Walther P99 Training and Shooting Practice Pistols

Remarks:

For training and shooting practice purposes Walther produced a number of P99 models. These are available as P99 QA, P99 DAO and PPQ pistol variations. They come in the traditional colors of practice models: blue, red or green. These identification colors were chosen according to the guidelines "Kennfarben für Munition" (code colors for ammunition)

P99 "Red Gun"

identical with the original in construction details and in operating elements, but not a live weapon

P99 "Blue Gun"

identical with the original in construction details and in operating elements, but only a marker gun; there is, however, a "blue gun" as well, which is only a plastic cast of the original (without any function, but in the original dimensions)

P99 "Green Gun"

shooting practice pistol with laser technology

P99 cutaway pistols

demonstration pistols for authorities and other users, cutting surfaces marked with red paint, SNs prefixed with "S"

Picture 88: *Walther P99 cutaway pistol, SN S400, caliber 9 mm Luger*

Walther P99 Compact (P99 C) Walther/Ulm

Picture 89: *Walther P99 compact, SN FAF 0087, caliber 9 mm Luger, variation for Hong Kong police (trigger relocated 3 mm to the rear), only 100 pistols made*

Production:	03/2004 - to date
Caliber:	9 mm Luger (9 x 19 mm) - 10 rounds
Total Length:	6.6 in.
Height:	4.3 in.
Barrel Length:	3.5 in.
Weight:	21.3 oz.
Rifling:	6 grooves, right-twisted
Operation:	same as full size P99 pistols, i.e. locked blowback breech, hammerless, striker-fired; short recoil, tilting barrel (modified Browning system); double action/ single action trigger mechanism

Characteristics: as "full size" P99 pistols, but

- To date frames and slides only available as black/black variation
- Trigger system: to date available as P99 compact AS only; "DAO" and "QA" variations are no longer available; P99 DAO compact pistols are still produced for authorities on special request, however.
- Slide stop levers: optional either one short lever on left hand side of frame only or extended, ambidextrous levers on both sides
- Magazine release levers: optional either short, ambidextrous levers or extended, ambidextrous levers on both sides of trigger guard
- High contrast sights: adjustable rear sight with phosphorescent paint markings ("3-dot" system) or Tritium night sights
- Interchangeable backstraps (two different sizes)
- 10 or 15-shot magazines with extensions (three different sizes)

Walther P99 and P99 Compact Licensed Production

P99 RAD (Poland) – from 2008

Characteristics: as standard "third generation" P99 Q pistols, but

- To date available in caliber 9 mm Luger only
- Slide legend on right hand side: "RAD" with Polish Eagle and triangular "FB" logo
- Slide legend on left hand side: "F.B. Radom Poland"
- Slide serrations: nine coarse serrations at the rear of the slide (P99 Q eight), no serrations at the front
- Slide contours: "softline design", top more rounded, no slanting surfaces or corrugations – cf. P99 Q
- Trigger module: DAO or QA – see above
- SN range: from PW0001

Remarks:

Before 2003 there was some sort of cooperation between Walther and the Polish company "Fabryka Broni Lucznik" ("FB", Weapons Factory Luznik) in Radom/Poland. Walther/Ulm delivered P99 AS pistols to the Polish police and FB was part of this transaction.

FB had gained a good reputation as a weapon manufacturer before 1945 already with the Radom 9 mm pistol VIS wz. 35. From 1953 to 1976 FB produced a copy of the Walther air pistol LP53.

After preliminary negotiations a contract between Walther and FB was concluded in 2008, which gave FB the status of a licensee. P99 pistols should be produced for the Polish army. Walther committed themselves to deliver complete frame assemblies and all the minor parts. FB was obliged to manufacture the slides and barrels, to make the final assembly and proof the P99 RAD pistols.

Moreover, it was agreed that FB should do the marketing of the P99 AS, QA, DAO and P99 compact pistols in Poland.

In 2008 Walther/Ulm shipped three sample pistols (SN WP00001 – WP00004) to FB in Radom, which were there given the FB inscriptions: "RAD" with "FB" logo and the Polish Eagle. These three P99 pistols still had three additional, coarse slide serrations at the front of the slide and the additional middle trigger safety (integrated trigger release lever in the trigger). Both features were omitted in the later series production.

One can sometimes read that "RAD" stands for "Rapid Application Development", but this is doubtful, as that term is used in IT for software development. The "R" presumably stands for "Radom". The word "Lucznik" (archer) in the company name refers to the old company logo – an archer.

SW99 (P99 Smith & Wesson, USA) – 1999 to 2006

Characteristics: as standard "second generation" P99 pistols, but

- Available in calibers 9 mm Luger, .40 S&W and .45 ACP
- Slide legend on left hand side: "Smith & Wesson / Springfield MA USA" with model designation "SW99"
- Slide serrations: 14 serrations at the rear of the slide, 7 serrations at the front
- Slide contours: top less rounded, less slanting surfaces at the front to gain more weight and thus inertia for the more powerful calibers
- Walther/Ulm provided the complete frames with "S&W" logo on both sides, but the

Walther banner on the left hand side and the inscription "Frame Made in Germany" on the right hand side; front of trigger guard rounded, bulge-shaped frame around the short slide stop lever on left hand side of frame

- Short, ambidextrous magazine release levers
- "two tone" variation with black frame and nickel-plated slide available

Remarks:

In the late 1950s, soon after Walther/Ulm was allowed to produce "sharp" weapons again, they were approached by the legendary Samuel Cummings. He became the importer and licensee for the US market. Cummings owned "Interarmco" (later renamed "Interarms") in Alexandria/Virginia.

Amongst other weapons this company imported Walther PPs, PPKs, PPK/S, TPH pistols etc. They also had these pistol models produced under license with their company address on the right hand side of the slide – see there.

Earl Sheehan and his company "Earl's Repair Service" in Tewskbury/Massachusetts became the co-importer and provided a repair and parts service.

After the business relationship between Walther/Ulm and Interarms had ended Walther founded a US subsidiary "Carl Walther USA LLC" in the year 2000 – located first in Alexandria/Virginia and later in Springfield/Massachusetts.

Two years later already Walther started a new business relationship with Smith & Wesson in the same location. In the years before, Smith& Wesson had helped Walther/Umarex with the sales of gas and alarm pistols in the US.

Smith & Wesson not only produced Walther pistols under license (see there), but also manufactured their own variation of the Walther P99 as SW99.

This production was performed in a so-called "joint venture". Hereby Smith & Wesson produced mainly pistols in the "typical American" calibers .40 S&W and .45 ACP and less in caliber 9 mm Luger.

By the end of 1998 Walther/Ulm manufactured the first 20 sample pistols for S&W already with the "S&W" slide legend.

SW99 Variations

SW99

DA/SA trigger with "anti-stress" trigger module, corresponds with the Walther P99 AS

SW990 and SW990L

DAO trigger module, corresponds with the Walther P990; the SW990L is the "full size" version; the SW990 came as a "compact" pistol as well – corresponding with the later SW99c

SW99 QA

QA ("Quick Action") trigger module, corresponds with the Walther P99 QA

SW99c

compact variation of the SW99, corresponds with the Walther P99 compact; early SW99c pistols have no Weaver or Picatinny rail, however

Total production of Smith & Wesson SW99 pistols: about 65,000 pistols

Walther PPQ ("Polizeipistole Quick Defense") Walther/Ulm

Picture 90: *Walther PPQ, SN FAL 7539, caliber 9 mm Luger, produced by Walther/Ulm in 2012*

Production:	01/2012 - to date
Calibers:	9 mm Luger (9 x 19 mm) - 15 or 17 rounds 10 x 21 mm (= .40 S&W) - 12 or 14 rounds .45 ACP - 12 rounds
Total Length:	7.1 in. (S&W: 7.2 in)
Height:	5.3 in.
Barrel Length:	4.0 in. (S&W: 4.1 in)
Weight:	24.1 oz. (S&W: 24.9 oz.)
Rifling:	6 grooves, right-twisted
Operation:	locked blowback breech, hammerless, striker-fired; short recoil, tilting barrel (modified Browning system); double action/ single action trigger mechanism

Remarks:

The Walther PPQ pistol is a further development of the Walther P99 Q and basically belongs to the fourth generation of the P99 model range.

This model was developed for special forces use only. Therefore the Walther PPQ was designed according to the "Technische Richtlinien Pistole" (technical specifications for pistols) of the German police forces and the "Polizeitechnische Institut" ("PTI" – technological institute of the police) of the "Deutsche Hochschule der Polizei" ("DHdPol." – German police academy). Due to the differing trigger weight the PPQ could not yet be certified, however.

After the model PPQ M2 was issued, the standard PPQ was renamed PPQ Classic.

Characteristics: as Walther P99, but

- Frame as P99 Q, i.e. new design of frame: Picatinny rail (according to MIL-STD-1932), "Hi-Grip" surface with a sanded texture, built-in service chip, lanyard loop at the heel of the grip; model designation embossed on both sides of the frame (near the heel); shape of frame modified in the trigger guard and accessory rail area; wider trigger guard for shooting with gloves
- Slide as P99 Q, i.e. new design of slide: top more rounded, no slanting surfaces or corrugations (so-called "softline design"); seven coarse serrations at the front and eight at the rear of the slide; no de-cocking button in top of slide; model designation no longer applied by laser, but deeply engraved
- Trigger module: "Quick Defense" system with pre-cocked striker and short trigger reset travel permits rapid firing; trigger travel (7 mm) and trigger weight (25 Newtons) are the same from the first to the last shot
- Extended, ambidextrous slide stop lever
- Extended, ambidextrous magazine release levers
- High contrast sights: adjustable rear sight with phosphorescent paint markings ("3-dot" system) and interchangeable front sights (made of steel); Tritium night sights optional
- Safety system: no manual safety, but four "automatic" (shooter independent) safeties:

a) striker is locked in idle position by a spring-loaded bolt and unlocked only if the trigger is pulled; this serves as a drop safety as well

b) trigger safety blocks the trigger in the frame, if no pressure is exerted on the trigger; with an additional middle trigger safety (integrated release lever in trigger)

c) slide closing safety: disconnector makes sure that the trigger bar is disconnected, if the slide is not completely closed

- Metal finish: tenifer treated
- Available: longer, threaded barrel (4.6 in.) with a thread protector for silencer
- Two row magazines with 15 resp. 17 rounds capacity (17 round magazine with extension), body coated with "Anti-Friction-Coating" ("AFC")
- Walther PPQ pistol available as "PPQ Blue Gun", i.e. a plastic cast of the original (without any function, but in the original dimensions)

Walther PPQ Navy (from 2012)

Remarks:

Not only combat swimmer units ("Seals"), but also other navy troops suffered from the problem that their service pistols were either completely flooded or at least soaked with water. So Walther/Ulm modified the PPQ pistol for the use in and around water.

Characteristics: as standard PPQ pistols, but

- Striker mechanism modified by additional holes in the striker channel and in other parts of the slide for an efficient water drainage
- Stronger striker spring to compensate for the water resistance
- PPQ Navy (SD) with longer, threaded barrel (4.6 in.) and a thread protector (for the use with a silencer) available

Picture 91: *Walther PPQ M2, SN FAM 0843, caliber 9 mm Luger, "fifth generation P99", photo: Carl Walther/Ulm*

Walther PPQ M2 (from 12/2012)

Remarks:

Because of the newly designed frame the PPQ M2 pistol belongs to the fifth generation of the of the P99 model range.

Characteristics: as Walther PPQ, but

- Upper magazine release button on left hand side of slide (behind the trigger) instead of the ambidextrous levers integrated into the trigger guard; easily convertible to right hand side
- PPQ M2 Sport: available with 5" barrel, longer slide and two rows of cutouts in the top of the slide for weight reduction
- PPQ M2 Navy (SD) available with longer, threaded barrel (4.6 in.) and a thread protector (for the use with a silencer)
- Magazines: 15 rounds (standard) or 17 rounds (optional), not compatible with the ones for the PPQ Classic series

Remark:

From 2015 onwards the PPQ M2 is available on the US market in caliber .45 ACP as PPQ 45, too.

Walther PPQ M2 .22 l.r. Walther/Arnsberg (from 2014)

Remarks:

This small bore model was developed as a shooting practice pistol for law enforcement authorities and target shooters. It has the design features and operational elements of the big bore variation.

Characteristics: as Walther PPQ, but

- Slide made of anodized lightweight alloy
- .22 l.r. barrel: length either 4" or 5" (the latter with a slide extension at the front)
- Magazine capacity: 12 rounds
- PPQ M2 Tactical: threaded 4.6" barrel for use with a silencer

Walther P99, P99 Compact and PPQ Walther/Ulm, Use by Armies and Police Forces

The different variations of the Walther P99 were and are used by the following police and armed forces:

- Rheinland-Pfalz (police): 10 000 P99 Q (from 2008) – SNs prefixed with "RP"
- Nordrhein-Westfalen (police): 42 000 P99 DAO (from 2005) – SN prefixed with "NRW"
- Hamburg (police): 8000 P99 Q
- Bremen (police): 2000 P99 Q
- Schleswig-Holstein (police): 8000 P99 Q
- Thüringen (police special task force): P99 AS
- Netherlands (police): 45 000 P99 Q (delivery: 2013 – 2015)
- Spain (several police forces)
- Portugal (National Police and National Guard): about 1500 P99 AS
- Malaysia (Royal Malaysian Police): 18 000 P99 AS und QA
- Thailand (police): amongst others 32 pistols P99 with longer slides, some P99 C
- Poland (police, border guards and judicial authorities): from 2001 to 2005 purchase of 10 000 Walther/Ulm made P99AS pistols, from 2008 use of FB made Polish RAD P99 pistols – see there; a total of more than 100 000 P99 AS pistols in use by 2014
- Mexico (federal police): P99 AS
- Italy (amongst others the municipal police of Florence): P99 AS
- Canada (police forces of Quebec "SPVQ" and Montreal "SPVM")
- Finland: army (special forces, military police) P99 Q introduced as "Pistool 2003" (short: "PIST 2003"); police, border guard, customs 7800 P99 Q – from 2012 onwards
- Ireland (plain clothes police): P99 C
- Nottinghamshire/GB (special task forces): P99 DAO
- South Africa (South African Police – SAP): P99 AS with additional, ambidextrous manual safeties
- Australia (Police): P99 QA and P99C QA
- Iraq (police)
- Estonia (police and border guard): P99Q (from 2/2014)

Walther P22 Walther/Arnsberg

Picture 92: *Walther P22, "First Edition", SN AB 0010, caliber .22 l.r., produced by Walther/Arnsberg in 2001, 10th pistol out of 100*

Walther P22 SN ranges:

P22 (pre-series):	**from SN 001 001** (additional "M" for "Muster", i.e. sample)
P22 (4" barrel):	**from SN G 001 001**
P22 (5" barrel):	**from SN H 001 001**
P22 Q:	**from SN Z 001 001**

Production:	04/2001 - to date
Caliber:	.22 l.r. - 10 rounds
Total Length:	6.3 in. (target model: 7.9 in.)
Height:	4.5 in.
Barrel Length:	3.4 in. (target model: 5.0 in.)
Weight:	16.9 oz.
Rifling:	6 grooves, right-twisted
Operation:	unlocked blowback action with firing pin and exposed hammer, double action/ single action trigger mechanism

Characteristics:

- Slide legend on left hand side: "P22 / Carl Walther Ulm/Do. / Made in Germany" or "P22" (with Walther banner)
- Proof marks: Cologne proof house ("Federal Eagle over N" with "three bishops' hats" and year code)
- Frame: made of polymer plastic (reinforced by glass fibers) with embedded steel trigger/barrel carrier module; burled anti-slip surface; available in several colors: olive green ("military"), carbon colored, pink, green, sandy-colored ("desert"), different camouflage patterns, etc.
- Accessory rail (Picatinny style) at the front of the frame (for e.g. a laser sight module)
- Interchangeable backstraps (in two sizes)
- Slide: made of die-cast zinc
- Slide serrations: nine on both sides at the rear of the slide and six at the front

- Fixed barrel held in a barrel carrier module; barrel sleeve held by a muzzle nut; barrel can be interchanged using a special tool to undo the muzzle nut
- Barrel threaded for silencer
- Recoil spring below barrel on a guide rod
- Exposed, rounded hammer with five grooves
- Slide stop lever on left hand side of frame
- Short, ambidextrous magazine release levers ("paddle-type") on both sides of trigger guard
- Ambidextrous take down catch in the frame (in front of the trigger)
- Ejection to the top right hand side
- No signal pin or loaded chamber indicator, but a sighting hole above the chamber
- High contrast sights: adjustable rear sight with phosphorescent paint markings ("3-dot" system) and interchangeable front sights
- Trigger system: double action/single action – i.e. first shot in DA mode, subsequent shots in SA mode
- Safety system:

a) ambidextrous, manual safety levers in the rear of the slide; when in "safe" position the safety shaft covers the end of the firing pin, which then cannot be reached by the hammer; no de-cocking function
b) internal drop safety in the slide locks the firing pin, if the trigger is not pulled
c) magazine safety disconnects the trigger bar, if the magazine is released
d) special key for the trigger lock (inserted into an opening on right hand side of frame)

- One row magazine, capacity: 10 rounds
- "two tone" variation with black frame and nickel-plated slide available
- P22 Target: with 5" and slide extension at the front available

Remarks:

The Walther P22 pistol is not a designated personal defense weapon. So it should actually not be dealt with in this book. But this model is technologically located between the P99 and the later PK380. This justifies its place here.

The Walther P22 was developed especially for the US market. So it consciously conveys the optical impression of being part of the "P99 family". But quite like the later Walther PK380 the P22 is actually a technological step back towards the roots of the Walther model range: The manual, pivoting safety levers in the slide show design features of the PP/PPK and TPH models. The P22 pistol has a polymer plastic frame and a modular design, however – like the P99 pistols.

The intended group of buyers are not necessarily the target or competition shooters, but the typical "spare time shooters" looking for a "fun gun" for some informal plinking shooting.

From 2001 onwards the Walther P22 is manufactured in series production. This is, however, not performed by Walther/Ulm, but by Walther/Arnsberg in Nordrhein-Westfalen (North Rhine-Westfalia). Some parts of the P22 pistol are made using the so-called "metal injection molding" ("MIM") technology. This is a special method of production, which Walther/Arnsberg in cooperation with Umarex has practiced for some years and has gained a lot of experience with – especially when manufacturing alarm and gas weapons.

There was a pre-series of 60 P22 first, which were marked the way Umarex usually marks their experimental pistols: with an "M" prefix (Walther/Ulm usually used a "V" prefix).

Walther P22 "First Edition" (from 4/2001)

The very first P22 pistols were issued as a limited edition (1000 pistols) in April 2001.

Picture 93: *Walther P22 Q, SN Z 128 121, caliber .22 l.r., produced by Walther/Arnsberg in 2012*

Characteristics: as later series production P22 pistols, but

- Slide legend on left hand side with addition "First Edition" (applied in white by laser)
- "First Edition" pistols came with both 3.4" and 5.0" barrels (the latter with slide extension)
- "two tone" variation with black frame and nickel-plated slide available
- Sold in luxury case with accessories

Walther P22 SD

As standard Walther P22, but with a threaded barrel for use with a silencer

Walther P22 Q

From 2012 onwards the model P22 is available as the P22 Q variation. This is more or less a "facelift" variation of the P22 pistol to adapt this model optically to the bigger PPQ.

Characteristics: as earlier P22 pistols, but

- Frame with model designation embossed on both sides (near the heel) – cf. PPQ pistols, "Hi-Grip" textured surface
- Slide with model designation no longer applied by laser, but deeply engraved (cf. PPQ pistols) without the "Q" in the model designation, however
- P22 Q Nickel: "two tone" variation with black frame and nickel-plated slide available
- P22 Q IL with integrated laser module available (for law enforcement authorities only)

Walther P22 Q USA

From 2009 to 2013 Smith & Wesson was in charge of the sales of the P22 resp. of the later P22 Q pistols on the US market. But on 1st January 2013 Walther Arms Inc., Fort Smith/Arkansas (the newly founded subsidiary of Walther/Ulm), took over the sales.

Walther PPS Walther/Ulm

Picture 94: *Walther PPS, SN AB 0013, caliber 9mm Luger, produced by Walther/Ulm in 2007, 13th pistol made as a test run in this caliber*

SN ranges:	**from AB 0001** (06/2007) in 9 mm Luger **from AB 7483** (12/2007) in .40 S&W **from AA 0001** (2008) in 9 x 21 mm
Production:	06/2007 - to date
Calibers:	9 mm Luger (9 x 19 mm) 9 x 21 mm (especially for export to Italy) 10 x 21 mm (= .40 S&W)
Total Length:	6.3 in.
Height:	4.4 in.
Barrel Length:	3.1 in. (.40 S&W: 4.1 in)
Weight:	21.5 oz. (.40 S&W: 22.9 oz.)
Rifling:	6 grooves, right-twisted
Operation:	locked blowback breech, hammerless, striker-fired; short recoil, tilting barrel (modified Browning system); double action/ single action trigger mechanism

Characteristics:

- Slide legend on left hand side: "PPS" (with Walther banner)
- Proof marks: Ulm proof house ("Federal Eagle over N" with "antler" and year code)
- Frame made of polymer plastic (reinforced by glass or carbon fibers) with inserted steel case for the slide and a ribbed/burled surface
- Accessory rail (Weaver) at the front of the frame (for e.g. a laser sight module)
- Interchangeable backstraps (in two sizes)
- Slide and minor parts made of tenifer treated or blued steel
- Eight slide serrations on both sides at the rear of the slide
- Tilting barrel (modified Browning system with a fixed, open cam), i.e. the upper part of the chamber serves as a locking block, which is supported in the slide against the front of the ejection port – cf. Walther P99

- Double recoil spring ("spring within the spring") below barrel on a guide rod
- Short slide stop lever on left hand side of frame with a bulge-shaped frame around it for protection
- Ambidextrous, extended magazine release lever integrated into trigger guard ("paddle type")
- Small, ambidextrous take down catch in the frame (in front of the trigger)
- Ejection to the top right hand side
- No loaded chamber indicator, but a sighting hole above the chamber
- Cocking indicator: red-dotted, rear end of striker serves as cocking indicator (cf. Walther Mod. 9)
- High contrast sights: adjustable rear sight with paint markings ("3-dot" system) and fixed front sight (made of steel)
- Trigger module removable from the frame
- Trigger system: "Quick Action" – i.e. striker spring is pre-tensioned about two thirds (6 mm); pulling the trigger ensures that the striker spring is fully tensioned by another 3 mm and the pistol is fired; all shots made with pre-cocked striker, i.e. with the same trigger weight and path from the first to the last shot;
- Safety system: no manual safety, but four "automatic" (shooter independent) safeties:

a) striker is locked in idle position by a spring-loaded bolt and unlocked only if the trigger is pulled; this serves as a drop safety as well

b) trigger safety blocks the trigger in the frame, if no pressure is exerted on the trigger; additional middle trigger safety (integrated release lever in trigger)

c) "QuickSafe" safety: The pre-cocked striker assembly is de-cocked simply by removing the backstrap. The removed backstrap therefore serves as a "key", too, as the pistol cannot be fired without it.

d) optional: magazine safety

- One row magazine, capacity: 6, 7 resp. 8 rounds (with extensions differing in length accordingly)
- "two tone" variation with black frame and nickel-plated slide available

Remarks:

The Walther PPS was designated "Polizei Pistole Schlank" (police pistol slim), as a worthy successor to the Walther police pistol PPK. In fact the PPS is only slightly larger than the PPK, but comes in much stronger calibers. Due to the trigger resp. the firing system and the operational elements the PPS is more a "smaller sister" of the P99 model series. This is a good reason for many users, who carry a P99 as a service or personal protection pistol, to use a compact PPS as their concealed "backup" weapon.

The problem is, however, that the Walther PPS cannot be certified according to the "Technische Richtlinien Pistole" (technical specifications for pistols) of the German police forces. The "Quick Action" trigger system does not comply with these specifications. Only police "Sondereinsatzkommandos" ("SEK", i.e. special task forces) are free to use non-certified weapons. So for them the PPS is an ideal "backup" pistol.

Hence the export market is of greater interest. The main target group [no pun intended!] are e.g. those buyers from the US market, who want to carry a big bore personal protection weapon with compact dimensions. This is the main reason, why the PPS is produced in caliber .40 S&W as well.

The first prototype PPS pistols were made In 2003. From June 2007 this model was manufactured in series production and in September 2007 it was launched onto the market.

Walther PPS "First Edition" (from 2007)

The very first PPs pistols were issued as a limited edition (1000 pistols) in 2007.

Picture 95: *Walther PPS, "First Edition", SN AA 0730, caliber 9 x 21 mm, produced by Walther/Ulm in 2009, 730th pistol made*

Characteristics: as later series production PPS pistols, but

- Slide legend on left hand side with addition "First Edition" (applied in white by laser)
- Frames made with a mottled silvery-anthracite surface by foil-coating, backstraps and magazine extensions are black
- Slide and operational elements made of black tenifer treated steel
- Three magazines in different sizes (6, 7 and 8 rounds)
- Sold in aluminum case with accessories
- SN range: AA 001 – AA 1000

Walther PPS "Police" (from 2007)

Characteristics: as series production PPS pistols, but

- Available in caliber 9 mm Luger only
- Steel sights with phosphorescent paint markings ("3-dot" system)
- Magazines: three magazines in different sizes (6, 7 and 8 rounds) are standard

Walther PPS Licensed Production

There is no real licensed PPS production. Two non-German companies import completely assembled PPS pistols and sell them only on their own markets.

- **PPS Radom**:

The Polish company "Fabryka Broni Lucznik" ("FB", Weapons Factory Luznik) in Radom imports the Walther PPS and applies the additional slide legend "F.B. Radom Poland" on the left hand side of the slide by laser.

The Polish, Estonian and other East European police forces show interest in procuring such "Polish" Walther PPS pistols as service pistols.

- **PPS Smith & Wesson**:

Smith & Wesson serves the US market with PPS pistols made by Walther/Ulm. In Ulm already these export pistols are roll-stamped with the legend "Smith & Wesson,

Springfield MA" on the right hand side of the slide.

From 1st January 2013 Walther Arms Inc., Fort Smith/Arkansas (the newly founded subsidiary of Walther/Ulm), took over the sales of the Walther PPS in the USA.

Walther P380 Walther/Arnsberg

Picture 96: *Walther PK380, caliber 9 mm kurz (= .380 ACP), produced by Walther/Arnsberg in 2009, photo: Carl Walther/Ulm*

SN ranges:	**from KP 001 001** (variation for European market) **from WB 001 001** (new variation for European market) **from PK 001 001** (variation for US market)
Production:	2013 - to date
Caliber:	9 mm kurz (9 x 17 mm = .380 ACP) - 8 rounds
Total Length:	6.5 in.
Height:	5.2 in.
Barrel Length:	3.7 in.
Weight:	19.2 oz.
Rifling:	6 grooves, right-twisted
Operation:	locked blowback breech; firing pin and exposed hammer; short recoil, tilting barrel (modified Browning system); double action/ single action trigger mechanism

Characteristics:

- Slide legend: "PK380" (with Walther banner) on both sides of slide
- Proof marks: Cologne proof house ("Federal Eagle over N" with three "bishops' hats" and year code)

- Frame made of polymer plastic (reinforced by glass or carbon fibers) with metal inserts and a ribbed surface
- Accessory rail (Weaver) at the front of the frame (for e.g. a laser sight module)
- Interchangeable backstraps
- Slide and minor parts made of blued or tenifer treated steel
- Eight slide serrations on both sides at the rear of the slide
- Tilting barrel (modified Browning system with a fixed, open cam), i.e. the upper part of the chamber serves as a locking block, which is supported in the slide against the front of the ejection port – cf. Walther P99
- Recoil spring below the barrel on a guide rod
- Partially exposed hammer with rounded spur and four grooves
- No slide stop lever
- Ambidextrous, short magazine release levers integrated into trigger guard "paddle-type")
- Ambidextrous take down catch in the frame (in front of the trigger)
- Ejection to the top right hand side
- No loaded chamber indicator, but a sighting hole above the extractor
- High contrast sights: adjustable rear sight with paint markings ("3-dot" system) and fixed front sight, tritium night sights optional
- Trigger system: double action/single action – i.e. first shot in DA mode, subsequent shots in SA mode
- Safety system:

a) ambidextrous, manual safety levers in the rear of the slide; when in "safe" position the safety shaft covers the end of the firing pin, which then cannot be reached by the hammer; no de-cocking function

b) internal drop safety in the slide locks the firing pin, if the trigger is not pulled

c) special key for the trigger lock (key inserted into an opening on right hand side of frame); key serves as a disassembly device, too (inserted into an opening on left hand side of frame)

- One row magazine, capacity: 8 rounds

Remarks:

The Walther PK380 pistol was developed especially for the needs of the US American market – quite like the Walther models P22 and PPS before. This may simply be concluded from the fact that this pistol is only produced in the caliber .380 ACP (9 mm kurz), which is traditionally very popular in the USA. So this caliber designation became even part of the model designation.

In one respect the PK380 is a technological step back towards the roots of the Walther model range: The manual, pivoting safety levers in the slide show design features of the PP/PPK and TPH range.

Some users – especially in the USA – keep asking for an inexpensive, "medium-format" pistol with manual safeties and an exposed hammer. So in the PK380 design some PP/PPK features could be resumed in combination with the new polymer technology and the robust Browning locking mechanism.

The result is a compact pistol in an effective caliber, which can easily be carried as a concealed weapon. So the Walther PK380 model finds itself in the tradition of the Walther PP/PPK and the P99 range.

An important intended group of buyers are senior and female shooters, who need a pistol which can be loaded and cocked with little effort. These users often find it difficult to handle a big bore pistol with its strong recoil spring.

From January 2009 Walther/Arnsberg started the series production of the PK380 pistols. In the years 2010 to 2012 Walther/Ulm had to help out in production though. Basically Walther/Arnsberg resp. Umarex is ideally equipped to manufacture such pistols. Some parts of the PK380 pistols are not made using traditional machining or even forging methods. These are made using the so-called "metal injection molding" ("MIM") technology. Walther/Arnsberg resp. Umarex are very experienced with this new technology, which they use especially in the production of alarm and gas weapons resp. the Walther P22.

From 2009 to 2013 Smith & Wesson was in charge of the sales of the PK380 pistols on the US market. But on 1st May 2013 Walther Arms Inc., Fort Smith/Arkansas (the newly founded subsidiary of Walther/Ulm), took over.

On 4th October 2012 Walther started a recall campaign for all PK380 pistols produced between June and September 2012 (SN PK 101 201 – PK 112 155) due to problems with the manual safeties. All PK380 pistols reworked in the factory were stamped with "Z" on the rear underside of the slide.

Walther PPX Walther/Ulm

Picture 97: *Walther PPX, caliber 9 mm Luger, produced by Walther/Ulm in 2013, photo: Carl Walther/Ulm*

Production:	2013 - to date
Calibers:	9 mm Luger (9 x 19 mm) - 16 rounds 10 x 21 mm (= .40 S&W) - 14 rounds
Total Length:	7.3 in.
Height:	5.6 in.
Barrel Length:	4.0 in. (in both calibers)
Weight:	26.4 oz.
Rifling:	6 grooves, right-twisted
Operation:	locked blowback breech; firing pin and shrouded hammer; short recoil, tilting barrel (modified Browning system); double action/ single action trigger mechanism

Characteristics:

- Slide legend on left hand side: "PPX" (with Walther banner) on left hand side of slide
- Proof marks: Ulm proof house ("Federal Eagle over N" with "antler" and year code)
- Frame made of polymer plastic (reinforced by glass or carbon fibers) with embedded metal rails and a "Hi-Grip" surface, lanyard loop at the heel, model designation on both sides of grip (near the heel)
- Accessory rail (Picatinny rail – "MIL-STD-1913") at the front of the frame (for e.g. a laser sight module)
- Underside of accessory rail with a small inserted plaque carrying the serial number and the proof marks
- No interchangeable backstraps
- Slide and minor parts made of tenifer treated or blued steel
- Slide serrations: six on both sides at the rear of the slide and five at the front
- Tilting barrel (modified Browning system with a fixed, open cam), i.e. the upper part of the chamber serves as a locking block, which is supported in the slide against the front of the ejection port – cf. Walther P99
- Barrel tenifer treated on inside, too
- Longer, threaded barrel (4.6 in.) with a thread protector (for the use with a silencer) available
- Recoil spring made of flat spring wire, below barrel on a guide rod
- Shrouded hammer flush with the rear of the slide
- No slide stop lever
- Upper magazine release button on left hand side of slide (behind the trigger) instead of the ambidextrous levers integrated into the trigger guard; easily convertible to right hand side
- Take down catch on left hand side of frame (in front of the trigger) – similar to P38/P1, P4 and P5
- Ejection to the top right hand side
- No loaded chamber indicator, but a sighting hole in the ejection port
- High contrast sights: interchangeable front sights; rear sights with paint markings ("3-dot" system) adjustable with special tool; tritium night sights optional
- Trigger module removable from the frame
- Trigger system: pre-cocked firing pin and short trigger reset (5 mm) permit rapid firing; trigger travel (7 mm) and trigger weight (29 Newton) are the same from the first to the last shot
- Safety system:

a) striker is locked in idle position by a spring-loaded bolt and unlocked only if the trigger is pulled; this serves as a drop safety as well

b) hammer is locked in pre-cocked position after the first shot - cf. safety catch of e.g. Walther P5

c) additional drop safety integrated into slide

- "two tone" variation with black frame and "stainless" slide available
- Two row magazine, capacity: 14 resp. 16 rounds
- SN range: within P99 serial number range (SNs starting with F…)

Remarks:

The Walther PPX pistol was developed as an inexpensive service pistol or commercial personal protection weapon. The main sales market will presumably be in the USA first, as this pistol model to date is not yet certified for the use by German authorities.

The Walther PPX pistol is a "hammer pistol" again – although the hammer cannot be cocked from the outside, as there is no hammer spur. Such a spur can easily be entangled in clothing – so at least it is said in adverts. On the other hand there has always been a demand for "hammer pistols" on the US market. So the shrouded hammer without a spur is a compromise.

Walther Pistols: Summary

Model	Production Period	SN range about	Total Production about
Model 1 Zella- St. Bl.	1910–1915?	1?–31 000	30 000 Pistols
Model 2 Zella- St. Bl.	1913–1915?	1 000–71 000	19 000 Pistols
Model 3 Zella- St. Bl.	1913–1915?	11 000–35 000	4 000 Pistols
Model 4 Zella-Mehlis	1913–1925/26?	20 000–493 700	351 000 Pistols
Model 5 Zella-Mehlis	1915–1923?	63 000–390 000	120 000 Pistols
Model 6 Zella-Mehlis	1915?–1917	1–1 038	1 000 Pistols
Model 7 Zella-Mehlis	1915–1919?	4 000– 84 000	45 000 Pistols
Model 8 Zella-Mehlis	1920/21–1943?	390 000–96 900A	145 700 Pistols
Model 9 Zella-Mehlis	1921–1943?	410 001–197 960N	248 000 Pistols
PP Zella-Mehlis	1929–1945	750 050–397 000P	540 000 Pistols
PPK Zella-Mehlis	1931–1944/45	754 900–431 097K	290 900 Pistols
PP Manurhin/Ulm	1952–1999	10 001–709 985	400 000 Pistols
PPK Manurhin/Ulm	1954–2003	100 001–814 332	633 000 Pistols
PPK-L Man./Ulm	1954–1982	500 001–533 640	33 640 Pistols
PPK/S Manurhin/Ulm	1968–1989	238 581–806 271	115 100 Pistols (until 1989)
PPK/E Ulm	2000–2010	001 001–?	6 800 Pistols
PP-Super Ulm	1974–1981	10 001–101 300	12 400 Pistols
TP Ulm	1961–1971	10 001–203 427	14 670 Pistols
TPH Ulm	1969–2000	250 001–305 137	55 137 Pistols
Mod. HP/Mod.P38	1937/38 –1945	1 001–25 800	24 800 Pistols
P.38 Zella-Mehlis Zero Series	1939–1940	01–013 720	13 695 Pistols
P.38 Zella-Mehlis	1940–1945	1–027 800	562 570 Pistols
P.38 Mauser	1942–1946	1–L-Block	326 350 Pistols
P.38 Spreewerk	1942–1945	1–00 100	265 900 Pistols
P38/P1 Ulm/Man.	1957–1997	001 001–473 201	473 200 Pistols
P38 K Ulm/Manurhin	1972–1981	348 513–502 595	2 653 Pistols

Model	Production Period	SN range about	Total Production about
P4 Ulm/Manurhin	1975–1981	600 001–607 419	7 413 Pistols
P5 Ulm	1978–1998	001 001–105 303	104 303 Pistols
P5 C Ulm	1987–1996	150 001–155 569	5 569 Pistols
P88 Ulm	1984–1992	001 001–010 894	9 894 Pistols
P88 C Ulm	1992–2000	100 001–107 344	7 367 Pistols
PP90 Ulm	1990	V 1522–V1535	7 Pistols
P99	1996–to date	001 001–?	
P99 C	2004–to date		
PPQ	2012–to date		
P22 Arnsberg	2001–to date	G/H/Z 001 001–?	
PPS Ulm	2007–to date	AB 0001–?	
PK380 Arnsb./Ulm	2009–to date	KP/WP/PK 001001-?	
PPX Ulm	2013–to date	P99 SN range ("F.....")	

Proof, Acceptance and Other Marks on Walther Pistols

"Deutsches Reich":

- **Commercial proof mark from 01/09/1911**: Crown over N

- **Commercial proof mark from 01/1040**: Eagle over N

- **Military proof mark from 1935**: Eagle over Swastika

- **Military proof mark from 1938**: Eagle over 359 (Walther only)

- **Military proof mark from 1940**: Eagle over WaA 359 (Walther only – see text)

- **Police acceptance stamp**:
  Eagle over Cross
  (with suffix letter C, F, K or L)

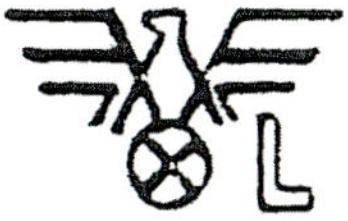

- **"Staatliches Materialprüfungsamt Dahlem"**: stylized "MPBD"
  (in P.38 grip plates until 1944 only)

Other markings until 1945

Walther, Zella-Mehlis production

- **D**eutsche **R**eichs**p**ost (alternative spelling "DRP") **D.R.P.**
- **R**eichs**j**ustiz **R.J.**
- **R**eichs**z**eug**m**eisterei **RZM**
- **R**eichs**f**inanz**v**erwaltung (with inventory no: "xxxx W" or "xxxx KW") **R.F.V.**
- **N**ational**s**ozialistisches **K**raftfahrer **K**orps **N.S.K.K.**
- Sturm**a**bteilung **SA**
- Schutz**s**taffel **SS**
- **R**eichs**s**icherheits**h**aupt**a**mt (SS) **RSHA**
- **R**eichs**b**ahn**d**irektion **R.B.D.**
- **R**eichs**r**undfunk**z**entrale **R.R.Z.**
- **P**olizei**d**irektion **M**ünchen **P.D.M.**
- Tiroler **St**andschützen**v**erband (Austria) **T.ST.V.**
- Danish state police ("**R**igs**pol**itie**t**") – after 1945, too **Rplt.**
- Norwegian police – after 1945, too **Politi**
- Polish army **"G"** over **"2"**

"Bundesrepublik Deutschland" (BRD):

- **Commercial proof mark from 1952**:
  Federal Eagle over N with proof house symbol (see below), from 1961 with the last three digits of year

- **Commercial proof mark from 1968**:
  Federal Eagle (new shape) over N, with proof house symbol and last three digits of year

- **Commercial proof mark from 1971**:
  Federal Eagle (stylized "thin" shape) over N, with proof house symbol and last three digits of year

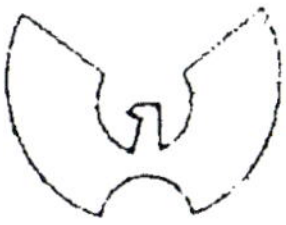

- **Commercial proof mark from 1973**:
  Federal Eagle ("fat" shape) over N, with proof house symbol and year of proof, the latter from 1977 coded (as described below)

Proof house symbols of the individual proof houses:

- Kiel (stylized nettle leaf)

- Hannover (Lower Saxony horse)

- Munich (Bavarian shield of diamonds, curved rim)

- Cologne (three "bishops' hats")

- Ulm ("antler")

- Berlin ("bear")

- Mellrichstadt (Bavarian shield of diamonds, straight rim)

- **Year codes** (from 1977):

A	B	C	D	E	F	G	H	J	K
0	1	2	3	4	5	6	7	8	9

Other Markings from 1945

Walther/Ulm and Manurhin production

- Proof and acceptance stamp of "Güteprüfdienstes des Bundes-Verteidigungsministeriums" ("Eagle with drooping wings", with numbers or "BWB" within the wings)

- Proof and acceptance stamp of "Güteprüfdienstes des Bundes-Verteidigungsministeriums" ("Eagle with drooping wings", as before, but with "BMI" within the wings)
- Property stamp of the "Bundesministeriums des Innern" (BMI procures weapons for the riot police forces, the state police, the border guard and secondary authorities, e.g. Office for the Protection of the Constitution): **BMI**
- Property stamp of the "Bundesfinanzministerium" (e.g. customs): **BUND**
- Property stamp of the "Bundeszollverwaltung" (alternative stamp for customs): **BZV**

- Property stamp of the "Deutschen Bundesbahn" (railway): **DB**
- Property stamp of the "Bundeswehr" (German army): **BW** (from 1977: **Bw**)
- "Bundesamt für Wehrtechnik und Beschaffung": **BWB**
- Maintenance marking "Bundeswehr" (with date): Maintenance center St. Wendel: **W 10.99** (Oct. 1999); Maintenance center Doberlug-Kirchhain: **DK 11.98** (Nov. 1998)
- Police Bremen: **Pol. Br.**
- Police Bremerhaven: **Pol. Brhv.**
- Police Rheinland-Pfalz: **R-P**
- Police Bavaria (state police): **LPBy**
- Dto. (later): **ByP**
- Police Niedersachsen (state police): **LPN**
- Dto. from 1970: **Nds** (in a square)
- Police Saarland: **S** (in a circle) or **PS**
- Police Hessia: **state emblem** ("rampant lion"), often with the abbreviation for the city, e.g. "DA" = Darmstadt)
- Police Baden-Württemberg: **state emblem** ("three stapled lions")
- Police Hamburg: **state emblem** ("castle gate with three towers)
- Police Schleswig-Holstein: **state emblem** ("two lions left hand side, nettle leaf right hand side")
- Police West Berlin ("Berlin Rosette"):

- Mark of the "Alliierten Kommandantur Berlin": **AKB**
- Police Sweden: **Tillhör Polisen**
- Police Norway: **Politi**
- Police Denmark ("Rigspolitiet") : **Rplt.** (with inventory no.)

German Democratic Republic (GDR):

- Commercial proof mark from 1945

(with Suhl proof house symbol and month/year, e.g. "660" = June 1960):

- Commercial proof mark from 1950

("Suhl Crown" over N, often with Suhl proof house symbol and month/year):

- Commercial proof mark from 1974

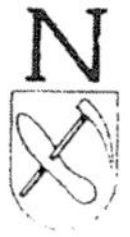

(N over Suhl proof house symbol):

Additional Markings:

- Stamp of "Abnahmebeauftragte" – see text ("A.B.9" in a wreath):

- Property stamp "Volkspolizei"

("VOPO star" with number of VOPO district – see list below):

Identification numbers of VOPO districts:

1 = Rostock
2 = Schwerin
3 = Neubrandenburg
4 = Potsdam
5 = Frankfurt/Oder
6 = Cottbus
7 = Magdeburg
8 = Halle
9 = Erfurt
10 = Gera
11 = Suhl
12 = Dresden
13 = Leipzig
14 = Karl-Marx-Stadt (Chemnitz)
15 = Berlin
16 = Wismut

Austrian Republic (2nd Republic)

- Commercial proof mark Vienna proof house: **NPv**
- Commercial proof mark Ferlach proof house: **NPf**
- "Gendarmeriebeschaffungsamt" (police procurement office): **GBA**
- "Heeresbeschaffungsamt" (army procurement office): **HBA**
- Property and acceptance stamp (Federal Eagle):

- Property stamp of the Austrian "Bundesheer": **BH**
- Property stamp of the Austrian "Bundeszoll" (federal customs): **Z**
- Property stamp of the Austrian "Sicherheitswache" (police): **SW**
- Property stamp of the Austrian "Justizwache" (judicial authorities): **JW**

Proof Codes of the Austrian Proof Houses - from 1982

- (first letter for the month, second and third letters for the last two digits of the year of proof)

Months:
E = Januar
L = Februar
N = März
B = April
S = Mai
Z = Juni
G = Juli
P = August
I = September
C = October
V = November
A = December

Code numbers:
W = 1
K = 2
R = 3
F = 4
M = 5
H = 6
Y = 7
T = 8
D = 9
O = 0

Walther/Ulm in-house factory markings:

- Quality control (pistols for the military and authorities):

- Dto. (commercial pistols): as before, but no dot in the middle
- Design modified (pistols for the military and authorities):

- Hardness test mark

- Dto.

- Letter prefix for pre-series/prototypes (Manurhin only): **W**
- Letter prefix for pre-series/prototypes (Walther only): **V**
- Additional letters for double assignments of serial number ranges: **D, R** oder **L**
- Marking on cutaway pistols "Ausschuss-Schnittmodell" (Walther only): **AS**

Bibliography

Hans Aicher, "Slimline aus Schwaben" [Walther PPS], in: DWJ 11/2007, p. 49 - 53.

Jan Balcar, Die Spreewerk P.38, Pilsen/Czech. Rep. 2009.

Jack H. Buchert, "P.38", in: Gun Collector's Digest, Digest Books, Northfield, Ill., USA, no year

Prof. Warren H. Buxton, The P.38 Pistol, Los Alamos, New Mexico, Vol. 1 (1978), Vol. 2 (1985), Vol. 3 (1990).

Prof. Warren H. Buxton, "The Gray Ghost P.38", in: Guns of the World, Petersen Publ. Co., Los Angeles, California, USA, 1972.

A.J. R. Cormack, Famous Pistols and Handguns, Profile Publications, Windsor, 1977.

Ulrich Eichstädt and Wolfram Kriegleder, "Carl der Große", in: Visier 5/1991, p. 6 - 23.

Ulrich Eichstädt, "Von der Stange" [Test Walther P99 AS], in: Visier 3/2012.

Edward C. Ezell, Handguns of the World, Arms and Armour Press, London-Melbourne, 1981.

Edmund Fasnacht, "Walther PPK-L, Fliegerpistole Eidgenössisches Ordonnanz Modell 1965", in: Waffenfreund (VdW), Issue 1/1996.

Robert Field, "Walther P.88 compact", in: DWJ 8/1992, p. 1224-1231.

Gene Gangarosa Jr. , P.38 Automatic Pistol, South Hackensack, N.J., 1993.

Joachim Görtz, "Produktionszahlen der P.38 1939 - 1942", in: DWJ 9/1982.
Joachim Görtz & Dieter H. Marschall, SS-Walther PP/PPK 1933 - 1944, self-published, 1999.

Joachim Görtz, "Walthers Pech-Marie: Das Modell 4 im Kriege", in: Waffen-Digest 2000, p. 147 - 152.

Thomas Greven, "Die Walther PPK-Rplt", in: Waffenfreund (VdW), issue 3/1994.

Daniel Guthanss, "Kampfpreis" [Walther PPX], in: DWJ 03/2013, p. 58 -60.

E.J. Hoffschmidt, Know Your Walther P.38 Pistols, Blacksmith, Chino Valley, Arizona 1975.

E.J. Hoffschmidt, Know Your Walther PP & PPK Pistols, Blacksmith, Chino Valley, Arizona 1974.

Ian Vernon Hoog, Die Deutschen Pistolen und Revolver 1871 - 1945, Motorbuch-Verlag, Stuttgart 1972.

Siegfried S. Hübner, "Die neue Combatpistole PP-Super", in: DWJ 3/1974, p. 262 - 269.

Martin Hugo und Martin Benz, "Dienstmädchen" (P.38) in: Visier, 8/1991, p. 88 -95.

Manfred Kersten, Walther - Eine deutsche Legende, Weispfennig-Vlg., Wuppertal 1997.

Manfred Kersten, "Die P99 von Walther", in: Waffenfreund (VdW), issue 3/1996.

Manfred Kersten, "Die Walther PP90", in: Waffenfreund (VdW), issue 4/1997.

Manfred Kersten, Dr. David Schiller u. Ulrich Eichstädt, Walther - Eine deutsche Erfolgsgeschichte, (2 Bd.), Weispfennig-Vlg., Wuppertal 2012.

Dr. Stefan Klein, "Die Walther-Pistolen TP und TPH", unpublished manuscript, 2013.

Martin Klesper, "Verkleinert", insert barrels Walther PP and PPK, in: DWJ 2/2014, p. 82 - 85.

Tom Knox (Ed.), several issues of AUTO MAG (monthly publication of National Pistol Collectors' Association NAPCA) St. Louis, Missouri.

Klaus-Peter König, Das Große Buch der Faustfeuerwaffen, Motorbuch-Verl., Stuttgart 1980.

Klaus-Peter König, "Walther P38 lang", in: IWS 4/1988.

Wolfgang Kräusslich, "Runderneuert" [Walther P99], in: DWJ 9/2004.

Max Krieg, "Walther P4", in: SWM 2/1988.

Alexander Krutzek, "Unfreiwillige Bande, FN und die P.38", in DWJ 7/1998, p. 1030 - 1034.

Alexander Krutzek, "Geburtstagskind - Die Pistole P.38 feiert ihren 60. Geburtstag", in: DWJ 2/1998, p. 249 - 254.

Norbert Kuch, "Sag niemals nie" [history of PP/PPK pistols], in: Visier 3/1993.

Dieter H. Marschall, "Lizenziert und abgekupfert, Walther PP/PPK-Kopien und Nachbauten" (part I - IV), in: DWJ 1 - 4/1998.

Dieter H. Marschall, "In eigener Regie - eine Walther PP aus Ulm für polizeiliche Linksschützen?", in: DWJ 8/1998, p. 1230 - 1231.

Dieter H. Marschall, "Carl Walther, Ulm", in: DWJ 4/2002.

Dieter H. Marschall, "Fußstapfen, Kurzbiographie Karl-Heinz Walther", in: DWJ 11/2003.

Dieter H. Marschall, "Weniger ist mehr?" [signal pins PP and PPK pistols], in: DWJ 10/2000.

Dieter H. Marschall, "Die Walther-Pistolen P38 und P1 der Bundeswehr", in: DWJ 9/2000.

Dieter H. Marschall, "Genesis, Teil 1" [Walther P5], in: DWJ 7/2000.

Dieter H. Marschall, "Genesis, Teil 2" [Walther P5], in: DWJ 8/2000.

Dieter H. Marschall, "Weit gereist" [PP for Bulgaria], in: DWJ 1/2000.

Dieter H. Marschall, "Nischenexistenz" [P21/PPK Bundeswehr], in: DWJ 2/2001.

Dieter H. Marschall & Dr. Geoffrey Sturgess, "Ermittelt, die Entwicklung der Walther PP", in: DWJ 10/2001.

Dieter H. Marschall, "Shooting Star" [Walther PP90], in: DWJ 6/2002.

Dieter H. Marschall, "Falsche Fünfziger" [Walther Model 5], in: DWJ 3/2003.

Dieter H. Marschall, "Walther PPK a la Chine" [Chinese copy of PPK], in: DWJ 5/2003.

Dieter H. Marschall, "Fremde Federn" [Walther Mod. 4 copies], in: DWJ 10/2003.

Dieter H. Marschall, "Schräubleskunde" [Walther PP made of plexiglass], in: DWJ 3/2004.

Dieter H. Marschall & Vjatscheslaw Artemenko, "Robuste Baltin, Vorbild für die Baltietz-Pistole war die Walther PP", in: DWJ 10/2004, p. 56 - 57.

Dieter H. Marschall, "Ende der Dienstzeit" [Walther P38], in: DWJ 3/2005.

Dieter H. Marschall, "Et cetera PP" [Walther PP copies], in: DWJ 1/2006.

Dieter H. Marschall, Die Faustfeuerwaffen der bewaffneten Organe der SBZ/DDR, Journal Verlag Schwend, 2001.

Dieter H. Marschall, "Bisher unbekannter PPK-Prototyp", in: DWJ 4/2012, p. 72-74.

Per Mathisen, Walthers P.38 Pistol, Mjoendalen/Norway 1998.

Per Mathisen, Walthers P.38 Pistol Spreewerk, Mjoendalen/Norway 2000.

Per Mathisen, Magazine Spotters Guide to P.38 Pistols, Mjoendalen/Norway 2002.

Per Mathisen, Walthers P.38 Pistol in Norway, Mjoendalen/Norway 2005.

Per Mathisen, A Pistol for Sweden - the m/39, Mjoendalen/Norway 2015.

Earl Mount, "Walther Model 9 Variations", internet contribution: http://unblinkingeye.com/Guns/WM9V (3rd Jan. 2012)

Gerhard Ortmeier, "Ustascha?", in: DWJ 03/2013, p. 64 - 67.

Karl R. Pawlas, Waffen Revue, issues nos. 102 - 110, Journal-Verlag, Schwäb. Hall, 1996 - 98.

S. Peelen, "P.38-Patente Zur Entwicklungsgeschichte der Walther P.38", in: DWJ 11/1975, p. 1218 - 1221.

James Rankin, Walther Pistolen 1908 - 1983 (German ed.), Stocker-Schmid Verl., Zürich-Dietikon, 1983.

James Rankin, Walther PP und PPK 1929 - 1945, 1982.

James L. Rankin, "Walther & Manurhin, The Story of the Postwar PP and PPK Pistols", in: Gun Collector's Digest, 3rd ed. DBI Books, Northfield, Jll., USA, no year.

Orvel Reichert, "P38 Pistol Production: 1939 - 1953", in: AUTO MAG, vol. XXXI, Issue 1, p. 19, St. Louis, Missouri, 4/1998.

Rudolf Schaar, "Taschenpistolen: Die Walther TPH", in: DWJ, 7/1972, p. 680 - 687.

Karl Schäfer, "Die Walther-Armeepistolen": in: Waffen Digest '96, Stocker-Schmid-Verlag, Zürich-Dietikon, 1995.

Karl Schäfer, "Einzelstück, eine Walther-Kniegelenkverschluss-Pistole?", in: DWJ 2/1988, p. 248 - 253.

Karl Schäfer, "Markenzeichen von Waffenherstellern", in: Schuss und Waffe 2, Dietikon, 1985.

Karl Schäfer, "Die Walther Pistole Modell PP für die iranische Polizei", in: DWJ, 3/1982, p. 294 - 298 and 4/1982, p. 432 - 435.

Arpad von Schalscha-Ehrenfeld, "Für Litauen", in DWJ 04/2015, p. 70 - 73.

Klaus Schinmeyer, "Die Selbstladepistole Walther P38 K", in: DWJ 9/1974, p. 945 - 948.

Jürgen Schönebeck, "Die Pistolen Walther PP, PPK und ihre Varianten", in: DWJ 8/1988, p. 1119 - 1122, Issue 9, p. 1248 - 1251, Issue 10, p. 1398 - 1400, Issue 11, p. 1540 - 1545.

Jürgen Schönebeck, "Aufbereitet" [PPK Zella-Mehlis/Ulm], in: DWJ 6/2003.

Jürgen Schönebeck, "Die Ulmer Walther P38/P1" in DWJ 6/1986, p. 676 - 681.

Matthias Schörmal, "Zertifizierung von Polizeipistolen", in: DWJ 10/2000.

Matthias Schörmal, "Der Renner" [Walther P22], in: DWJ 7/2001.

Matthias Schörmal, "Schnelle Nummer" [Walther P99 QA], in: DWJ 1/2001.

Walther Schulz, "Geburtstag", in: DWJ 5/2006.

Dave Shike, "Late Date ac Frame P.38s", contribution to P.38-Forum (USA) March 2013.

Jürg Siegenthaler, "Walther Modell 2 oder Modell 5?", SWM 6/2007, p. 18 - 21.

W.H. Smith, Small Arms of the World, Stackpole Books, Harrisburg, Penn. 1960.

R.A. von Steindler, "Selbstladepistole Indian Arms P380", in: DWJ 10/1977.

Dieter Stiefel, "Die denkende Schusswaffe" [Walther Mod. 9], in: DWJ, 12/1979.

Egon Thiel, "Ausländische Walther-Pistolen in West-Berlin", unpublished manuscript, Meerbusch 1995.

Prof. Frank Thielemann, Kurzwaffen der Firma